Essentials of User Interface Design

Essentials of User Interface Design

Alvin Albuero De Luna

ARCLER
PRESS

www.arclerpress.com

Essentials of User Interface Design

Alvin Albuero De Luna

Arcler Press
224 Shoreacres Road
Burlington, ON L7L 2H2
Canada
www.arclerpress.com
Email: orders@arclereducation.com

ISBN: 978-1-77469-136-6 (Hardcover)

ABOUT THE AUTHOR

Alvin Albuero De Luna is an instructor at a Premier University in the Province of Laguna, Philippines - the Laguna State Polytechnic University (LSPU). He finished his Bachelor's degree in Information Technology at STI College and took his Master of Science in Information Technology at LSPU. He isnhandling Programming Languages, Cyber Security, Discrete Mathematics, CAD, and other Computer related courses under the College of Computer Studies

TABLE OF CONTENTS

LIST OF FIGURES

LIST OF ABBREVIATIONS

AI	Artificial Intelligence
AJAX	Asynchronous JavaScript and XML
API	Application Program Interface
AR	Augmented Reality
ATM	Automated Teller Machine
CEOs	Chief Executive Officer
CLI	Command-Line Interfaces
CMOs	Chief Medical Officer
CSS	Cascading Style Sheets
GDI	Graphics Device Interface
GUI	Graphical User Interface
HCI	Human-Computer Interface
HTML	Hypertext Markup Language
IBM	International Business Machines
IoT	Internet of Things
NIA	National Institute on Aging
NLP	Natural Language Processing
OLED	Organic Light-Emitting Diodes
OS	Operating Systems
PC	Personal Computer
RAD	Rapid Application Development
UI	User Interface
UID	User Interface Design
UML	Unified Modelling Language
UNIX	Uniplexed Information Computing System
UX	User Experience
VR	Virtual Reality

VUI	Voice User Interface
WIMP	Windows, Icons, Menus, and Pointers
XAML	Extensible Application Markup Language
XML	Extensible Markup Language
XSLT	eXtensible Stylesheet Language

PREFACE

This book takes the readers through different stages of the user interface, from tracing its history to bridging the gap from user requirements to design to UIDs built for programmers. It also focuses on the visual interface design, to defining the patterns used as tools for the user interface. It depicts the importance of communication and languages to further defining the challenges and opportunities and the role of UIDs in the coming year.

This book sheds light on several aspects of the user interface, such as the history, the golden rules, the levels, how it connects users to designs, and the different methods to do so. It also discusses the interaction between a human and a computer, along with defining the prototyping, the color theory, to patterns. The later chapters give an insight into how visual communication is vital to make conversational interface design. The problems and opportunities and the use of UI in the future.

The first chapter stresses the basic overview of the user interface so that the readers are clear about the philosophies behind that form the utmost basics in the field. This chapter will also emphasize the history and sheds light on several aspects of user interface such as the history, its different types, how humanity came back to computing, the golden rules of designs used in a user interface, and the various levels of designs marking from visceral, behavioral and then going towards reflective level.

The second chapter takes the readers through the concepts of user requirements and the bridge built between user requirements and designs, between designers and engineers, designs and developments along with defining the problems faced in creating these bridges and how they can be overcome by maintaining communication and applying other methods. This chapter will provide highlights on the various key problems that are faced during bridging, the limitations and restrictions which come with bridging.

The third chapter deals with the user interface design for programmers, along with defining the various user interface principles. It signifies the human and computer interaction, and tells about the prototyping in the user interface. It mentions the usable design, the UI design process, and the importance of user interface tools for programmers. It also tells about the evaluation of user interfaces.

The fourth chapter introduces the readers to visual design in aesthetic and strategic design. The different visual elements in UI design, the five fundamental elements of interface design, the essential user interface design tips, the principles of visual design. This chapter also denotes the applying color theory to digital displays. It's designing for senior citizens and creating a UX design style guide.

The fifth chapter throws light on how patterns are used as tools for UID and how design patterns are considered as such powerful designs. This chapter also illustrates the user experience design patterns, from defining the importance of UX patterns to dark patterns that are used in a user interface.

The sixth chapter takes the readers through the concept of languages and communications, the characteristics of language. The readers are then told about how communication is associated with visuals. It explains the conversational interface designs, how a visual language is built, and then tells the reader why should they invest in a visual language.

The seventh chapter explains the challenges and opportunities that prevail in a user interface, the basics of UID, and the literature review. It also lays out the challenges that are faced by UX designers. This chapter also emphasizes the biggest challenge in design, the user interface challenges, and common problems with the UX process.

The last chapter of this book sheds light on the future of user interface design, the current state of UI design, the establishment of voice user interface and accessibility, how artificial intelligence will disrupt UI design, the trends, and prophecies of UI design. It also compares the past, present, and future of UI. This chapter also mentions the evolution of interfaces, along with describing the future UI trends, then the impacts of UID on everyday life, the future of UX and UI design. It tells the reader about the brain, that is, the computer interface, then the two types of interfaces, telepresence, and lastly, the modern UI 2020.

This book has been designed to suit the knowledge and pursuit of the researcher and scholars and to empower them with various dimensions of User interface design from its history to how it is used in the present time and eventually what it will bring in the coming years so that they are updated with the information. I hope that the readers find the book explanatory and insightful and that this book is referred by scholars across various fields.

CHAPTER 1

THE USER INTERFACE: AN INTRODUCTION AND OVERVIEW

CONTENTS

This chapter explains the basic significance behind the concept or ideology of the user interface. This chapter also explains the history of the user interface, as well as it also provides highlights on bringing humanity back to computing. This chapter also sheds light on the various types of user interfaces. The chapter on user interface addresses the four golden rules that play a significant role in user interface design. This chapter also mentions the three levels of design that have been used in user interface design.

1.1. INTRODUCTION

A user interface is also referred to as UI or, in simple words, 'interface.' This means that a person controls a software application or even a hardware device. If there is a good user interface that provides a user-friendly experience and this further allows the user to interact with the software or hardware in a natural as well as intuitive way.

Around every software program have GUI that is a graphical user interface. In other words, it means that the program includes graphical controls, which can be selected by the user by using either a mouse or keyboard. A typical GUI that is Graphical User Interface of a software program includes a menu bar, windows, toolbar, buttons, and other controls.

The two operating systems, namely Macintosh and Windows, have different user interfaces. On the other hand, they tend to share a lot of the same things, including desktop, windows, icons, etc., such kind of common elements make it possible for users to use either of the two operating systems.

There is no need to relearn the interface again and again. Similarly, some programs named word processors and Web browsers all have similar interfaces and tend to provide a consistent user experience across various and multiple programs.

There are many hardware devices that include a user interface, and also, typically, it is not complex like the software interface. Consider a common example of a hardware device along with a user interface, and that is a remote control. A TV remote is a typical example that has a numeric keypad, volume and channel buttons, an input selector, power and mute buttons and also, some other buttons which tend to perform several different functions.

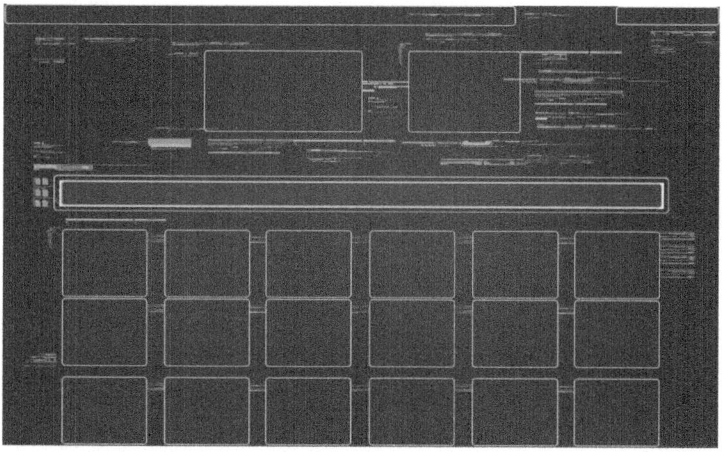

Figure 1.1: An illustration of the user interface.

Source: Image by Pixabay.

TV remote contains a set of buttons. This set of buttons and how they are laid out on the controller tends to make up the user interface. Some other devices, including digital cameras, stereo systems, and audio mixing consoles, also have a user interface.

On the one hand, the user interface can be designed for either hardware or software, and on the other hand, it can be designed for a combination of both as well. Let us consider an example, and that is to control a software program. In this, the user needs to use a keyboard and mouse, and these have their user interface.

Similarly, to control a digital camera, the user needs to navigate through the on-screen menus, and that is a software interface. Without considering the application being used, the main aim of a good user interface is to be user-friendly. Otherwise, it would become frustrating to use a device that does not work the way the user wants it to.

1.2. THE HISTORY OF USER INTERFACES

There was a time when Apple's new device, the Macintosh was kicking off a revolution in personal computing, and the reason was its graphical user interface that is GUI and the mouse. These were the two innovations that helped democratize computing because they made computers understandable as well as approachable for the average consumer.

Considering the time period of the 1990s and early 2000s, this period continued innovation stalled. But with the coming 2010s, a UI renaissance started resulting in a number of interactions which was powerfully disruptive and more of a human form.

To understand if we are continuing to become digitized, it is important to have a look at how humans have interacted with computing to date and also how these innovations have created the foundation for the newest forms of interaction.

1.2.1. 1960s–1980s

In the late 20th century, the keyboard further dominated human interactions with computing technology. There was no existence of touch interfaces, voice control, and some similar technologies. Yet, the keyboard was considered to be a vast evolution over the punch card. The punch card was used to program the early computers in the period of 1940s and 1950s.

1.2.2. 1984

In this year, Steve Jobs visited Xerox Parc, which was a famous event covered. There he found inspiration in the form of a GUI and mouse. Additionally, these two UI innovations created a seismic event in technology adoption. Apple sold approximately 1 million Macintoshes by the year 1988. There were some companies named IBM, Compaq, and some others which quickly followed with their computer mice.

This wave was followed by Microsoft. They introduced Windows 1.0 in the year 1985, then Windows 3.1, in the year 1992, where it began to feel the GUI tail wind.

1.2.3. 1994–1997

At the beginning of the 1990s, the laptop computer began to overtake the desktop. There were also increasing changes taking place along with the same in the mouse or keyboard interface. This was the time when Apple began to implement the trackballs and trackpads into its PowerBook laptops. On the other hand, IBM introduced pointing sticks into its laptops.

A new device was also becoming famous when computing continued miniaturizing from the desktop to the portable PC. This device was named the PalmPilot. With the handheld form factor, Palm introduced a new user interface named the stylus. This user interface worked with its touchscreen, and an alternative alphabet which was referred to as "Graffiti."

Along with this, the voice also started to become a form of interaction in the year 1997. This was the year when the introduction of Dragon NaturallySpeaking was taking place. Dragon (and later Nuance, which acquired Dragon in the year 2000) sold a large number of around million copies of its voice interaction software, although it would not be until the introduction of today's generation of voice assistants. In today's era, the voice would truly start to be considered as a form of computing interaction.

1.2.4. Early 2000s

Most of the time, the UI development was increasing during the period of early 2000s, when the improvement in the already-established devices and tools was being done:

- The introduction of the first commercially popular optical mouse (which was non-trackable) was done by Apple, and this was referred to as the Pro Mouse. Following this, they introduced later on the Mighty Mouse.

This added touch as well as pressure sensitivity as an upgrade to the previous interface, which was based on buttons. A multitouch Magic Mouse followed next. This further allowed users to have interactions in more advanced ways by using combinations of many fingers.

- In addition to this, Apple introduced the scroll wheel. Their user interface was based on touch along with the rollout of iPod music devices in the year 2001. With time, the scroll wheel became even more touch-sensitive which enabled Apple to remove all external buttons.

1.2.5. 2007–2010

In the late 2000s, people witnessed a big leap forward in UI that is user interface design, which was again led by Apple. Touch interfaces took off dramatically along with the introduction of the iPhone in the year 2007 and the iPad in the year 2010. These were the devices that had multipoint capacity touch. This, in turn, enabled users to interact with digital content in some new ways.

On the other hand, Apple was not the one who invented this UI (it came from Finger works, and it was acquired in the year 2005 to provide this). The company certainly democratized it, along with other smartphones as well as tablet manufacturers who all adopting forms of touch over the following years.

1.2.6. 2011–Present

Along with the time, when touch was becoming a ubiquitous part of the way humans interact with digital content today (smartphones and tablets as well as kiosks, ATMs, appliances, and much more), there was the development of revitalization invoice as a user interface form.

In the year 2011, the voice assistants began to gain steam along with Siri voice assistants, which is provided by Apple. From that time, humans also observed the introduction of Google Now (in the year 2012) and Alexa devices provided by Amazon (in the year 2014). Such services and devices depend on data as well as content assets that were acquired by such platforms to fulfill user requests.

Therefore, when people ask Siri for direction ns, it can quickly leverage Apple Maps so that it can provide a routing. Or when people ask an Amazon Echo (which is powered by Alexa) to play a song or read an Audible book, Alexa tends to draw on those Amazon assets so that it can playback the content.

There are several experiences enabled by a new generation which are considered to be virtual and augmented reality devices. These experiences add another dimension to user interfaces. Consumers, as well as enterprises, are just starting with new forms of interactivity in architecture, gaming, real estate, social networking, and engineering, including a few examples too.

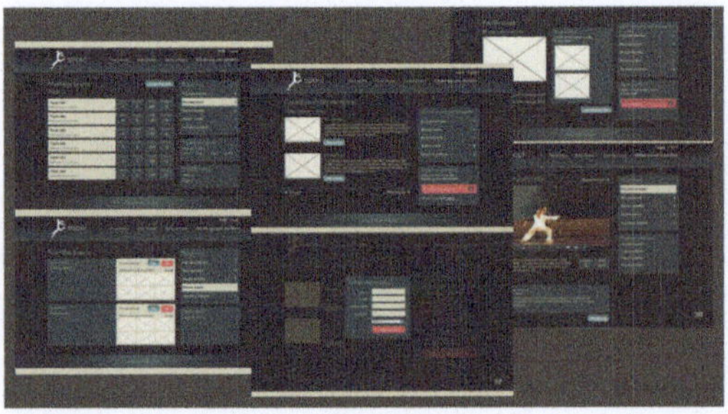

Figure 1.2: User interface design exploration.

Source: Image by Flickr.

In today's era, this means wearing a large device that is mounted on the head and using a mouse or gaming controller to interact with virtual objects. But as haptic devices (which allow for the perception as well as manipulation of objects using touch), reach some new levels of performance as well as miniaturization, anyone can imagine being able to touch, run, walk and interact with the virtual worlds in the natural way similar to that in which one interacts with the real world.

1.3. BRINGING HUMANITY BACK TO COMPUTING

Without considering the history that has just been discussed before, it can be understood that UI design is heading in a clear direction, and that is toward organic forms of interactivity which are native to biology.

Considering the previous times, it can be observed that the keyboards, mice, and trackpads were just hacks that are meant to close the gap between spoken language as well as touch along with the computing surfaces.

Nowadays, the era is on the edge of a return to much more natural interfaces which involve fingers, voices, and bodies in space. This has become possible by advances in the network infrastructures, computational power, and cloud-based computing storage, and aggregation of the data which is required to teach machines. This is done to understand as well as interpret human interactions.

Many interesting possibilities are opened by having these advancements of all sorts and these possibilities include:

- Visiting a pastime or place with being able to interact with the environment. One example can be, visiting the Court of Henry VIII and speaking to the king and also, with his courtiers.
- Attending the college graduation of a family member virtually, with being able to interact with the people around in real-time.
- Hosting a meeting in a virtual environment that has attendees from several different countries. Each participant speaks in the first language of the person, whereas the others hear real-time translation into their first language.

1.4. TYPES OF USER INTERFACE

The communication that takes place between a user and a computer is said to be two–way. Among different jobs, one of them is of the operating system, and that is to provide a user interface. This will eventually help humans to communicate with the hardware which makes up a computer. When a piece of software is bought, then it will also contain a user interface, to have access to the software and use it.

Data and instructions are given by a user to a computer, and the computer will give the information to the user back. How a computer and a user tend to communicate is referred to as the interface. Many alternative terms are there do as to describe this. One more common term that s being used is the HCI that refers to Human-Computer Interface.

If the interface is being fully described, then it is important to talk about the input devices, the software interface, and the output devices. There are five various types of an interface that can come with an operating system and these are mentioned below:

- GUI that is graphical user interfaces;
- CLI that is command-line interfaces;
- Form-based interfaces;
- Menu-based interfaces;
- Natural language interfaces.

Given below is the diagram that can summarize these five types of interface.

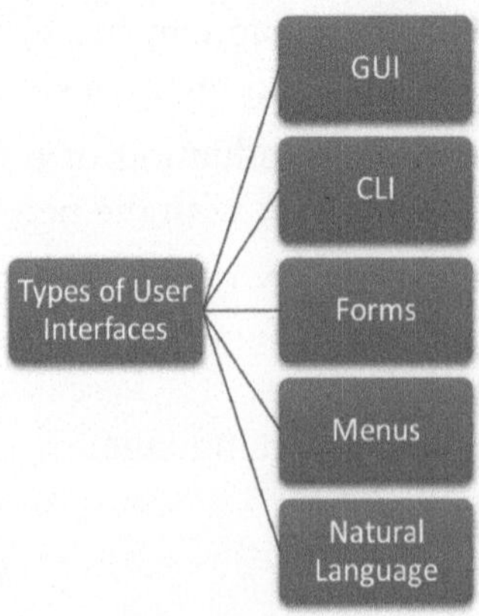

Figure 1.3: *Types of user interface.*

1.4.1. Graphical User Interfaces (GUI)

Those interfaces which are graphical are referred to as GUI that is Graphical User Interfaces. It is also termed as WIMP Interfaces, i.e., Windows, Icons, Menus, and Pointer. Typically, it can be expected that these types of interfaces are available in multi-tasking environments or in applications software which includes a considerable degree of complexity. A multi-tasking environment is an environment in which user tends to open and use more than one piece of software simultaneously.

Figure 1.4: An illustration of GUI icon.

Source: Image by Wikimedia commons.

Whenever anyone used Windows, Word or Office application or Explorer, or Firefox (to surf the internet), that user has used a GUI several times. Every such application has its window which is opened up into and the user can open up more than one window at a time (and thus, more than one application can be used).

At one time, only one application is active. Considering Windows, for example, one is active because the active window has a bright blue bar at the top of the window which is in opposition to a dimmed blue bar. In addition to this, there are icons too which the user can click on to access the tools fast in the application.

There are drop-down menus that make sure that the user does not have a large number of options continuously on display which can take up room on the screen. Usually, the pointer is a mouse or a figure on touch – screens, although the graphic designers as well as engineers, most of the time, prefer to use a tablet and stylus to point. The combination of a mouse or tablet or stylus tends to make sure that the user can select quickly instead of having to use a keyboard that is slower and prone to mistakes.

While summarizing, typically, it can be expected to find the following in a GUI or WIMP user interface:

- **A window for each open application:** Several windows can be opened simultaneously, but only one window can be active at a single time. There can be some way to indicate the one is active (that can be done by making the bar at the top of the active window bright blue).
- **Menus and icons:** Selection of the available functions can be done in one of two ways. These ways include using pop-up menus or drop-down menus or by clicking on icons. In simpler terms, an icon is said to be a small picture that tends to represent a specific function, and clicking on it selects that function.
- **A pointing device:** this is the device that is used to make selections. Typically, it is a mouse, a graphics tablet, and pen, or a finger on touch screens. The keyboards are used to navigate through the application, which is minimized because this task takes time.

Some companies make several various applications which try to be similar to the interface in each application. This, in turn, helps the users who are familiar with one application, to quickly select a new application that is designed by the same company.

Let's consider an example: When Excel is learned, then the learner didn't have to learn the way to open, close, and save a file, change the font and even how a picture can be inserted, and much more. This is because the learner was probably already familiar with Word and Excel as well; this is because both were developed by Microsoft.

1.4.2. Command Line Interface

A CLI that is Command Line Interface tends to require a user to type in commands from a list of allowable commands. Let's suppose that any user wants to back up a file which is termed donkey.doc to a folder (directory) called animals on the floppy disk. In a GUI (Graphical User Interface), the user will open the file manager, then click on the file that is to be saved and drag it to the folder named as animals on the floppy disk. Any user any human can do that.

Another example that can be taken is when the user wants to do the equivalent in DOS. DOS has a command-line interface. For this, the user needs to know how to construct the command to copy a file from one place to another place.

While using this type of interface, it can take a long time to learn, and that is not intuitive. For the users who are not experienced, it can be frustrating to use this type of interface whereas, for the users who are experienced, this type of interface can be very powerful. The reason behind this statement is that the command line interface provides many commands which can get a user very close to the workings of the components of any computer system.

There are many commands which can manipulate the hardware as well as software in a computer system in a way that cannot be done simply by using a GUI. In other terms, it can be said that there are tasks where the user has to use a command-line interface to carry them out. Some good examples include UNIX and DOS of operating systems which use such kind of interface.

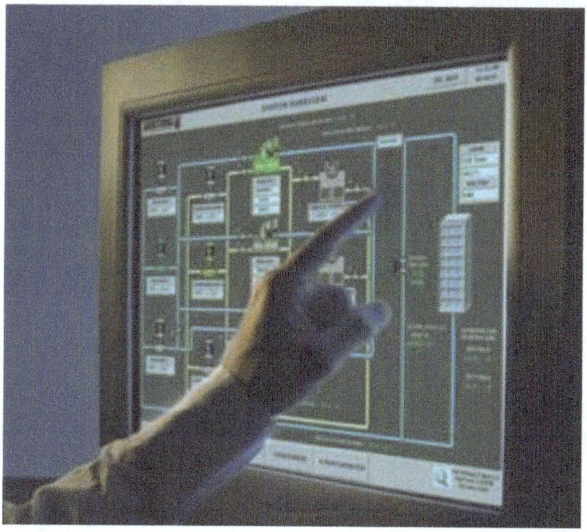

Figure 1.5: An illustration of HMI graphical user interface.

Source: Image by Wikimedia commons.

In the command-line interface, the typical users are technicians and network managers. They need to perform some setup tasks and system tasks. Such tasks can be done only by using such type of interface.

1.4.3. Forms

Some operating systems are designed only for business purposes where employees are required to enter lots of information. Let's consider a form that is paper-based, and then a person is asked to fill it in, that is for the membership of a club or an application for a driving license. The things to be written in the form are highly directed.

Several instructions are given to help the person while filling the form, including boxes where the person writes or selects information from choices as well as boxes where the person can simply tick one of a selection.

A form-based on software interface on a computer is considered to be much similar to an interface that is based on paper. It is expected that the input that is put into the computer is predictable. If a large number of forms-based interfaces are used, then the user starts to observe some common characteristics.

1. Field names, names next to a place are there, where the information must be entered. The places where information is being entered in by the user are referred to as 'response fields.'
2. There are some other types of response fields too, which include radio buttons and drop-down selectors.
3. The cursor tabs from one response field to the next automatically. This in turn guides the user logically by using the form, and it ensures that all the information required is collected.
4. As entry of data is done, it is then 'validated.' Validation tends to attempt to make sure that only entry of sensible data is done into the system and also, that the data which is not sensible is rejected. The validation helps in making sure that the data which is entered into any system maintains its consistency. In other words, any data which is stored is only of the format that is expected in a specific field. Data can be validated by using several methods. These types of methods include:

- A range checks.
- A data input mask.
- A character length check.
- A presence checks.
- Using check digits.
- Getting the user to select from a list using combo boxes or look-up tables.

5. If it becomes necessary, then the input can be changed or even canceled.
6. When an 'OK' button, ENTER, or something similar to that is pressed, then the data is finally entered into the system.
7. There is some sort of HELP facility that is available.
8. Some options are not displayed on the main screen. This is done to avoid cluttering up the form. The less commonly needed facilities can be accessed via a selection button that connects to a separate screen.

The interfaces which are based on forms are very suitable for any kind of application in which predictable pieces of information are entered into the computer. There can be many examples including:

- Someone who is recording responses to questions while telephone questionnaire is taking place;
- Someone who is taking telephone orders for a product like a CD;
- Someone who is entering in details of people who want to apply for a credit card;
- Someone who is buying something online;
- Someone who is applying to join a club or open a free email account on the Internet.

All of these activities that are mentioned above can be done with the help of an interface that is based on form. The reason is the same that the predictable information will be asked by the user or by the organization that is based on the web over and over again for either each order or questionnaire or application.

1.4.4. Menus

There are some operating systems that are designed with a menu-based user interface. The menu-based user interfaces are considered to be ideal for some situations. Situations like where the IT skills of the user cannot be guaranteed or in situations that tend to require selections that are to be made from a large number of options or in situations that require very fast selection. The user of a system that uses the menu-based interface will be presented with a limited number of options on the screen.

Once a selection has been done, then, the user is presented with a sub-menu. Moreover, this gives them many other options. They make another selection and further, this may be presented with a further sub-menu. This tends to continue unless the user gains the ability to select exactly what they want from the choices which are eventually displayed on the screen. Let's consider an example of a menu-based screen that can be found at a tourist office.

If there is a tourist who may not have any IT skills then this could be presented with a screen with nine buttons on it. These screens include theatres, cinemas, pubs, and trains while considering them as an example. They will then touch the touch screen in the area of the buttons so that they can make a selection. For example, if they selected 'Cinemas,' then they would be presented with a sub-menu.

This can also look like another menu-based screen having six buttons on it, one for each cinema in the area. After that, if they selected one of those, then they would be presented along with the films which are currently showing and the times on which they are. Such type of user interface is considered as simple as the user can get. The user does not need any kind of computer skills to access the wealth of information on a system similar to this.

Let's consider a factory where workers are working in a noisy as well as a dirty environment. The workers cannot want to be fiddling around with the keyboards like typing in commands. Instead of this, they can have a menu-based interface. This in turn will allow them to quickly find the option which they wanted. To select that option, they have to simply touch on a touch screen.

Usually, the fast-food outlets have a till which is made up of simple selections for the cashier so that he or she can choose from that. Sometimes, he or she can do this by using words on each button, and sometimes he or she can use pictures too. Further, this makes it easy for a cashier to take an order. A little training is required for them and it is said to be one way in which the fast-food outlets make the jobs in their restaurants low-skill. They do this so that they can then keep wages low.

1.4.5. Natural Language

This is the type of interface that requires the user to enter responses to the questions that are being asked by the computer. The questions are displayed on the VDU and the answers are entered by using the keyboard.

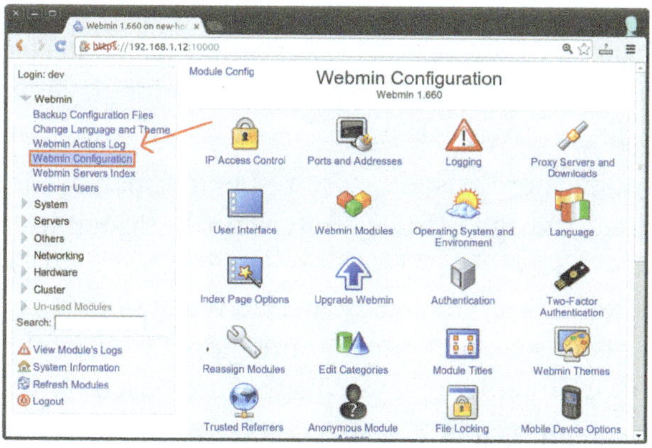

Figure 1.6: An illustration of GUI-based Linux server management.

Source: Image by Flickr.

This type of interface is referred to as a 'natural language' interface. The reason for this name is that the computer as well as the user tends to appear holding a conversation. Let's take an example. Imagine a 'save file' request is initiated by the user. The conversation can further go like this:

***USER*: Save file**

***COMP*: What is the file name?**

***USER*: Chapter1.txt**

***COMP*: What folder?**

***USER*: User_Guide**

***COMP*: File already exists. Overwrite?**

***USER*: Yes**

***COMP*: Done.**

Such type of interface can be found on data entry terminals and some other types of dumb terminals which are linked to a network (where non-expert users are guided by the computer through the not so simple tasks which are required to be performed by them).

1.5. THE 4 GOLDEN RULES OF UI DESIGN

The UI that is User Interface is said to be an important part of any software product. When it is done in a good manner, then the users do not even notice it. When it is done poorly, then the users cannot get past it to use a product efficiently. Most designers tend to follow interface design principles to increase the chances of success while creating user interfaces. Principles related to interface design tend to represent the high–level concepts which are used for guiding software design.

Various principles are based on Ben Shneiderman's The Eight Golden Rules of Interface Design, Bruce Tognazzini's Principles of Interaction Design, and Jakob Nielsen's 10 Usability Heuristics for UI Design. Most of these principles can be applied to any interactive systems that are traditional GUI environments (e.g., desktop or mobile apps, websites) and some non-GUI interfaces (e.g., interaction systems that are based on voice).

There are some UI design principles that have been mentioned below:

- Place users in control of the interface;
- Make it comfortable to interact with a product;
- Decrease cognitive load;
- Make user interfaces consistent.

1.5.1. Place Users in Control of the Interface

The UIs which good tend to have a sense of control in their users. This, in turn, keeps the users in control, making them comfortable. By this, they will learn quickly and also, gain a fast sense of mastery.

(i) Make Actions Reversible – Be Forgiving

This is the rule which means that the user is required to always be able to quickly back-track the thing that they are doing. This in turn allows the users to explore the product without the constant fear of failure. When in case, a user knows that the errors can be easily undone, then this tends to encourage the exploration of unfamiliar options. On the other hand, when a user needs to be extremely careful while performing any action, then it leads to a slower exploration and nerve-racking experience, which is not wanted by anyone.

There are some most common GUIs in which the users have the 'Undo/Redo' option and that includes text and graphics editors. When a text is written or graphics is created, then the 'Undo' lets users make changes and go back through the changes that were made by the user. The option of 'Redo' makes the user able to undo the undo. This means that once the user goes back a few steps, then they become able to move forward through their changes again.

The option of 'Undo' can consider being extremely helpful while the users choose the system function by mistake. In such a case, the undo function is considered to be an emergency exit. This in turn allows the users to leave the unwanted state. There can be one good example for such emergency exits, and that can be the Gmail's notification message having an undo option when the users delete an email accidentally.

(ii) Create an Easy-To-Navigate Interface

The navigation needs to be clear as well as self–evident. Users need to be able to enjoy while exploring the interface of any of the software products. Despite this, the complex B2B products, which are full of features should not intimidate the users so that they have a fear to press a button.

Any good UI tends to put the users in their comfort zone. This can be done by providing some context of where they are, where they have been and where they can go further.

- **Predictability**: The users are required to provide some cues which can further help them in predicting the result of an action. Any user is required to never be wondering about what should be pressed to do a task or what is any button for.
- **Provide visual cues**: for users, the visual cues are considered as reminders. This, in turn, allows the users to navigate easily through the interface, which can be done by providing points of reference along with their movement through a product interface.

The users are given an immediate view of where they are in the interface, which can be done by page titles, highlights for currently selected navigation options, and other visual aids give. Any users are required to never keep in mind that where he or she is or how did he or she get to this screen.

(iii) Provide Informative Feedback – Be Acknowledging

Typically, the feedback is related to the points of action for each user action, the system is required to show a meaningful as well as clear reaction. A system having feedback for each action tends to help the users in achieving their goals without any hurdle.

The UI design is required to consider the nature of the interaction. Taking frequent actions into account, the response can be modest. Let's consider an example. When the users tend to interact with an interactive object (like a button), it is important to provide some indication that acknowledgment of action has been done. This can be considered as something simple in comparison to a button changing color whenever pressed (the change notifies the user about the interaction). The lack of such kind of feedback tends to force the users to double-check while looking for their intended actions which have been performed.

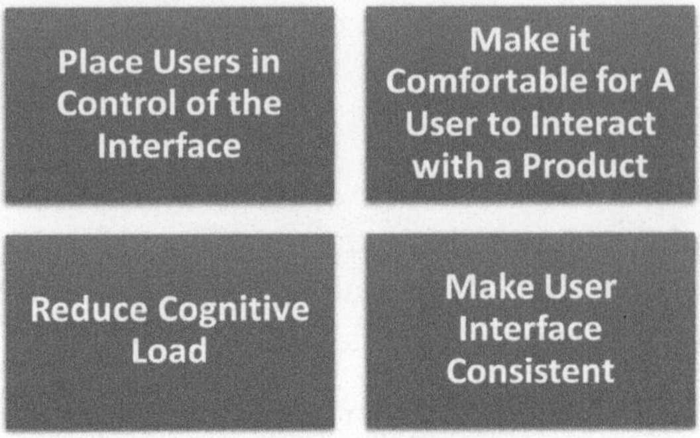

Figure 1.7: Four golden rules of user interface design.

Considering the infrequent and significant actions, the response is required to be more substantial. Let's take an example, while filling a password field in the signup form, UI (specifically a good UI) can inform users about the requirements for their password.

(iv) Show the Visibility of System Status

The users are said to be much more forgiving whenever they have information related to what is going on and are provided periodic feedback about the status of the process going on. The visibility of system status is important whenever the users initiate an action that takes some time for a computer to complete that.

The users do not like this thing that he or she is left seeing nothing on the screen when something is going on in the app. The progress indicators are used, and this is considered to be one of the subtle aspects of UI design that has a major impact on the comfort as well as enjoyment of users.

Good UI can provide comfort to the users, and this can be done by showing progress while a task is being completed by the system. Dropbox helps in indicating the status of a document upload and the status includes the current progress and the time that is being left.

(v) Accommodate Users with Different Skill Levels

The users who have different types of skill levels are required to have the ability to interact with a product at various levels. It is important to not sacrifice expert users for novice or casual users for an easy interface-to-use. Rather than this, it is important to try to design as per the needs of a large number and different kinds of users, therefore, it does not matter if the user is an expert or a newbie.

Features such as tutorials and explanations must be added, and that is extremely helpful for novice users. But it is important to make sure that the experienced users can skip this part. Once the users become familiar with a product, they will further look for shortcuts to increase their speed while performing some commonly-used actions. It is important to provide fact paths for experienced users, and this can be done by enabling them to use many shortcuts.

1.5.2. Make it Comfortable for a User to interact with a Product

(i) Eliminate all elements that are not helping your users

The interfaces are not required to contain some information that is relevant or rarely required. Not-so-relevant information tends to introduce noise in UI (user interface), then it competed with the appropriate information and diminished its relative visibility.

The interfaces should be simplified by removing elements or content which are not necessary and do not provide any support user tasks. The UI must be designed in such a way that all the information that is presented on the screen will be valuable and also, relevant. Every element must be examined and evaluated based on the value that it delivers to users.

There can be a good example for this, and that can be of an app that follows the less is more approach which is done by avoiding overloading the interface having content or features is iA Writer. The interface that is used in the iA Writer app is said to be a clean typing sheet having no distractions. Furthermore, it allows the users to focus on what they are writing and hides everything else.

(ii) Don't ask users for data they've already entered

The users must not be forced to have to repeat data that they have entered previously. Users become annoyed easily by tedious data entry sequences, especially whenever they have provided all the information that is required earlier. A good UI tends to perform a maximum of work when a minimum amount of information is required from the users.

(iii) Avoid jargon and system-oriented terms

While designing a product, it is important to use such a language that can be easily read and understood. The systems are required to speak the language of the user having words, phrases, and concepts that are familiar to the user instead of jargon or terms that are system-oriented.

(iv) Apply Fitts's law to interactive elements

The Fitts law states that 'the time to acquire a target refers to a function of the distance to and size of the target.' In other words, it is better to design large targets for important functions (big buttons are said to be easier to interact with).

In addition to this, it is important to remember that the time that is required to acquire many targets is the sum of the time to acquire each. Therefore, while working on UI design, it is important to reduce distances and increase target sizes and also, reduce the total number of targets with which the users must interact to complete a given task. this must be done to increase the efficiency of interaction.

(v) Design accessible interfaces

When products are designed, then it is important to keep in mind that a well–designed product is accessible to a user of all abilities, which includes those having blindness, low-vision, hearing impairments, motor impairments, or cognitive impairments. A good UI is accessible UI because when the product's accessibility is improved, then it tends to enhance the usability for all groups of users.

The color is considered to be one of the elements of an interface that has a strong impact on accessibility. People tend to perceive color in a different manner. Some users can see a full range of colors but also, some people can only see a limited range of colors. Around 10% of men and 1% of women have some kind of color blindness.

When the designing of interfaces is done, then it is better to avoid considering color as the only way to convey the information. In any phase of time, color is required to convey the information in the interface them other cues must be used to convey the information to those who are unable to see colors.

(vi) Use real-world metaphors

When the metaphors are used in UI design then this allows the users to create a link in between the real world and digital experiences. The real-world metaphors tend to empower the users. These tend to allow the users to transfer existing knowledge related to how things must be looked at or worked. Most of the time, metaphors are used to make the unfamiliar familiar.

The recycle bin must be used on your desktop as this holds the deleted files. This can be considered as an example that it is not a real trash bin but is represented visually in a way that in turn helps the user to understand the concept more easily.

Good metaphors tend to generate a strong link to past experiences from the real world in the minds of the users. On Macs, the recycle bin icon is considered similar to an actual bin and it represents whether it has files in it or not.

While a metaphor is chosen for UI, then the one is selected which will enable users to grasp the finest details of the conceptual model. Let's consider an example, when someone is asked for credit card details for processing payment, then a person can reference a real-world physical card considering it as an example.

(vii) Engineer for Errors

While considering the user journey, the errors are inadvertent. When the bad error handling is paired with useless error messages, then this can fill users with frustration, and this can lead to abandoning the app. On the other hand, a well–crafted error message can convert a moment of frustration into a moment of conversion. An effective error message is considered to be a combination of explicit error notifications altogether with hints to solve the problem.

In comparison to writing good error messages, it is better to have a UI design that tends to prevent a problem from occurring in the first place.

Elimination of the error-prone conditions must be tried or even can be checked. This in turn presents users with a confirmation dialog before the time when they commit to the action. An example can be taken of Gmail. Gmail prompts the user when they forget to insert an attachment.

While trying to prevent users from making those errors in the first place, the best designs tend to have an excellent error recovery. In Gmail, the error prevention shows a pop–up if the users forget to insert an attachment after referencing one.

(viii) Protect a user's work

It is important to make sure that the users never lose their work. The users should not lose their work which can be a result of an error on their side (like accidentally refresh a web page that has a form having user input), a problem with an internet connection, a system error, or any other such reason other than those reasons which cannot be avoided completely such as an unexpected power loss.

1.5.3. Reduce Cognitive Load

The quantity of mental processing power which is required to use a product is called cognitive load. It would be better to avoid making users either think or work too hard while using the product.

(i) Chunking for sequences of information or actions

The theory of chunking was brought by psychologist George Miller in the year 1956. Miller said that the working memory of a human could handle seven-plus/minus two "chunks" of information for processing any type of information.

While organizing and grouping any items, this rule is used. For example, if any UI forces the user to enter their phone numbers having no normal spacing in between, then the result will show a lot of incorrectly captured data. A cluster of ten or more numbers cannot be scanned to discover errors or incorrectly – captured data. This is a reason why phone numbers are split up into smaller fragments.

(ii) Reduce the number of actions required to complete a task

While designing a user interface, an attempt is made to lessen the total number of actions by the user to achieve the goal. The three-click – rule is always worth remembering in which a user will obtain all the information regarding a product in not more than three clicks of a mouse.

(iii) Recognition over recall

Jakob Nielsen provided usability heuristics, one of which advises promoting recognition over recall in the UI design. Recognizing anything is better than recalling to recognize something that involves more cues in the brain.

In user interfaces, the designers promote recognition which can be done by making information and functionality noticeable and easily accessible. Tooltips and context-sensitive details are some examples of visual aids, which can help support users while recognizing information.

(iv) Promote visual clarity

To quickly recognize or find the information which the user is looking forward to using the interface even more efficiently, a good visual organization helps in improving usability and legibility.

(v) When designing layouts:

- Providing too much information on the screen should be avoided. A grid system design should be constructed to avoid visual clutter.
- The principle of 'form follows function should be remembered to make things like they are working.
- The general principles of a content organization must be implemented, which can include grouping the same items, numbering the items, and further, using the headings as well as prompt text.

1.5.4. Make User Interfaces Consistent

Consistency is considered an important property found in a good UI; the consistent design is intuitive. Consistency is also considered among the strongest contributors to usability and learnability. The idea of transferable knowledge is the main idea of consistency. This idea lets the users transfer the knowledge and skills that they have from one part to another of the User interface of an app.

(i) Visual Consistency (Style)

The integrity of products should never be questioned by the user. The product should be presented with the same colors, fonts, and icons. Make sure that the design system manager should be referred so that the visual styles of the product are not changed for any possible reason. For example, the submit buttons on the user interface site on one page should be similar to all the other pages.

Different styles of elements should be avoided on different pages of the site. Users are not required to keep in mind that whether a transformed button similar to this example means a similar thing.

(ii) Functional Consistency (Behavior)

Consistency of behavior is when the object is working in the same way throughout the user interface. The action of interface controls, like buttons and menu items, should be the same as that of a product. Surprises and changes are not always liked by users, because of this, they get easily frustrated as things don't work the way they want them to.

(iii) Consistent with user expectations

While using the apps/website users have certain expectations. The worst thing for the user while designing a product is contradicting a user's expectations. It is not important that what logical argument is provided for how something should work or even look. If a different way is required by the user, then a very hard time is spent while changing according to the user's expectations. If the advantages of the approach offered are not clear, the work should be modified according to the user.

- **Follow platform conventions**.

The product should be presented according to the platform guidelines, which are standard dedicated for products. With guidelines, it is assumed that individual interface elements can be understood by a user.

- **Don't reinvent patterns**.

There are proper solutions for most of the designs. These proper solutions are called patterns. Users are already familiar with proper patterns which become conventions. If these solutions are not taken into account and keep designing, the solution will create new challenges for users. The result in breaking design conventions can result in frustrating user experience such as the user will face usability problems which can be because the users are not familiar with it and not because of the solution that will be wrong.

- **Don't try to reinvent terminology**.

New terms should be avoided when users already know the words. There are certain expectations for naming because a lot of time is spending by users on other apps as well as on other sites. Confusion is created while using different words.

1.6. THREE LEVELS OF DESIGN (DONALD A. NORMAN)

Over the years, there have been technological advancements taking place. Despite all of such technological advancements, it has been quite different to understand the complex working of the human mind.

In a book named 'Emotional Design: Why we love (or hate) everyday things," its author Donald A. Norman tends to differentiate between three aspects (or in other words levels) of the emotional system. The emotional system is referred to as the sum of the parts which are responsible for emotion in the human mind, and these include visceral, behavioral, and reflective levels.

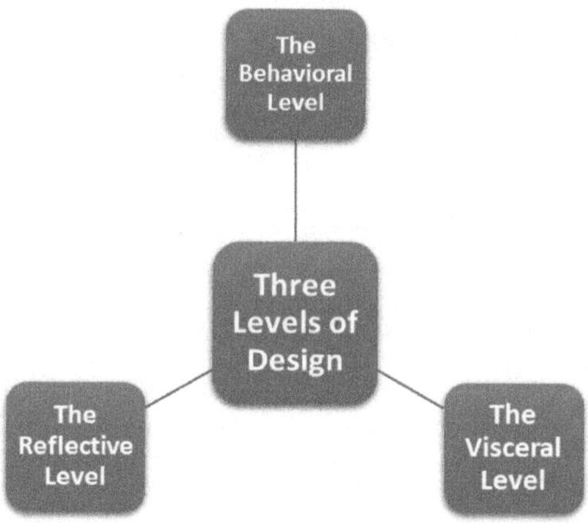

Figure 1.8: Three levels of design given by Donald A. Norman.

Each of such levels is entirely connected and interwoven in the emotional system. These, in turn, influence the design in its particular way (Norman, 2004).

1.6.1. The Visceral Level

Considering the most basic level of processing which is referred to as the Visceral level. This level is related to the basic mechanisms of protection of the effective system of the people who are responsible for quick judgments about the environment.

These judgments include checking is either good or bad, either safe or dangerous. It can further be said that the visceral level is directly linked with the motor system. This will give results, for example, for animals to fight or flee or even relax.

The unconscious thought is connected to the visceral system, which in turn allows the users to respond quickly as well as sub-consciously to the events. The responses provided give rise to the startle reflex for novel and also, unexpected events.

The reason behind this is that it is at a level where there is no conscious control on the person which tends to present immediate responses while producing an affective state. This affective state is not affected by either context or history of the situation. The Visceral level response tends to evaluate the event momentarily automatically, where any cause is not attributed.

An initial sensory experience further triggers a visceral reaction. This is considered to be that first impression which tends to set the mood ad initial framing for which the user will explore everything. Visceral reactions which are powerful and positive can set a positive context for each subsequent interaction.

It further makes the users forgive the faults down the line even if the initial experience was overwhelmingly positive. Moreover, this will encourage the positive socialization of the product (Baker, 2019).

Such a level of design refers to the perceptible qualities of the object and how they make the user feels. In addition to this, the Visceral designs focus to get inside the head of the user and tug at their emotions. This in turn improves the user experience or even serves some business interest (Komninos, 2017).

Ultimately, it can be said that this level has no connection with the way the product is to be used or how effective or understandable it is. It entirely focuses on attraction or repulsion. There are many great designers who use their aesthetic sensibilities to drive such visceral responses.

1.6.2. The Behavioral Level

Once, the user learns to completely act, when they want to perform it again, the only thing that is required to do is think about the goal itself, and most of the time the user subconsciously will act quite naturally. This is the situation that occurs when the user learns to drive a car and practice driving for weeks.

To drive a car, the person simply gets into the car, sits in the seat and everything else is done subconsciously in which almost no effort is required. The person is required to pay attention to the street while driving and in which direction he/she is going. But in such a case, the actual control of the muscles is under conscious perception, and his/her concentration is mainly on not hitting other cars. Therefore, his or her hands automatically get adjusted (Norman, 2004).

At this level, the actions, as well as analyses, are majorly sub-conscious. Although people are usually aware of their actions, most of the time, they are not aware of the details. Let us consider some examples.

When a person speaks then most of the time, he/she does not know what he/she is about to say until unless the conscious mind hears them uttering the words. In another case, when a person is playing a sport, then he/she is prepared for action, but their responses occur so fast for conscious control. This is where the behavioral level takes control.

The behavioral level is mainly relating to usability properly; this, in turn, refers to the practical as well as functional aspects of a product or any other useful object. The behavioral design tends to include usability, performance, product function, and also, the effectiveness of use.

Not similar to the visceral level (which has much more subjective as well as particular characteristics of every person), the behavioral level can be tested easily. An example can be as a person wan check the time taken by him/her to complete a task or the number of errors made by him/her while trying to act with a product.

There can be several other examples of experiences at the behavioral level which include the difficulty faced while typing on a small touch screen device, how easy it is to type on a computer keyboard, and the enjoyment that is felt by a gamer while he/she uses a well-designed game controller.

When there are products that enable the user to complete their goals with the minimum difficulty and by putting a little conscious effort, it is possible that the emotions thus occurred are the positive ones. Contrastingly, when there are products that restrict the user and force them to either translate or adjust their goals as per their limitations or even simply make them pay attention while using them, then the user tends to become more inclined to experience some of the negative emotions (Komninos, 2017).

The behavioral reactions which are powerful as well as positive, further allow users to feel a sense of empowerment, cultivate reliability and trust. This is done by creating a direct link between the actions and the expected value of the user. Along with this, these reactions also encourage repeat reactions, because people are more inclined to experience such delight again (Baker, 2019).

1.6.3. The Reflective Level

The reflective level is connected with conscious cognition which is not similar to the other two levels. Here the reasoning, as well as conscious decision making, is done. The refection at this level is said to be deep as well as slow, which is not similar to the unconscious thought.

The reflective level is mainly concerned with the user experience itself. it feels similar to the feeling after going through an immersive experience while using a product and how the user remembers the experience. This in turn determines whether or not the user want to try that experience one more time (Norman, 2004).

Usually, this happens after the events which have occurred along with a reflection on them. These events can evaluate the actions as well as the results. The highest levels of emotions come from the reflective level. This is where the causes are assigned and also, where the predictions of the future tend to occur.

The reflective reactions which are powerful and positive tend to encourage users to share their experiences with other users. Moreover, it helps in evoking a sense of pride and also in identifying while using a product that further extends beyond the product itself. Ultimately, the reflective emotional design tends to capture the meaning of the product, the ability to share the experience, the impact of thoughts, and the cultural impact (Baker, 2019).

1.7. CONCLUSION

In the conclusion of the chapter, the significance of the user interface has been discussed. This chapter also discussed the history of the user interface along with the concept of bringing humanity back to computing has been discussed. In this chapter, various types of several user interfaces have been discussed, such as graphical user interface, command-line interface, forms, menus, as well as natural language.

Towards the end of the chapter, four golden rules that play an important role in user interface design have been discussed, which emphasizes the place user in control of the interface, make it comfortable for a user to interact with the product, decreasing the cognitive load, and make user interfaces much more consistent. This chapter also discussed the three levels of design that include the visceral level, the behavioral level, and the reflective level.

REFERENCES

1. Asher, M., (2017). *The History of User Interfaces—And Where They are Heading*. [online] Adobe Blog. Available at: https://blog.adobe.com/en/publish/2017/07/18/a-brief-history-of-ui-and-whats-coming.html#gs.z0xhlu (accessed on 6 July 2021).
2. Babich, N., (2019). *The 4 Golden Rules of UI Design | Adobe XD Ideas*. [online] Ideas. Available at: https://xd.adobe.com/ideas/process/ui-design/4-golden-rules-ui-design/ (accessed on 6 July 2021).
3. Dutra, R., (2020). *Three Levels of Design (Donald A. Norman)*. [online] Medium. Available at: https://medium.com/swlh/three-levels-of-design-donald-a-norman-4f36a8db82d6 (accessed on 6 July 2021).
4. Techterms.com. (n.d). *User Interface Definition*. [online] Available at: https://techterms.com/definition/user_interface (accessed on 6 July 2021).
5. Theteacher.info. (n.d). *Types of User Interface*. [online] Available at: http://theteacher.info/index.php/systems-software/notes/4623-types-of-user-interface (accessed on 6 July 2021).

CHAPTER 2

USER INTERFACE DESIGN: BRIDGING THE GAP FROM USER REQUIREMENTS TO DESIGN

CONTENTS

This chapter is bridging the gap from user requirements to design in user interface design has been discussed in detail. The various ways to bridging the gap between design and development like teaching each other, working like one team, working together from the start have, being flexible testing, the assumptions have also been done in detail.

In this chapter, there is also a detailed discussion about various problems in bridging the gap has also been discussed. In this chapter, the ways to bridge the gap between design and development have been discussed in detail.

2.1. INTRODUCTION

When it comes to the subject of modern-day web and digital product development challenges, one has certainly mentioned the fact, the design-development divide or design-development overlap is one of the major issues. The expansion of this occurred, which likely resulted in most heads nodding as if to say they had just concluded that nothing would be said about it. In that case, after all, there must be a tool for that, correct. Is it eluded us or has it only not been found?

Instead of focusing on just using one tool, they can use many methods to fill the design productivity gap. However, the dilemma can be more difficult to solve than one would expect because of multiple potential solutions.

Working on software projects is difficult. Regardless of if they' whether, they are a planner, a product manager, or an engineer, they all have an essential part in the end user's final product. The path to producing outstanding software is not over until they have built the team; it is also important to maintain the working relationship with their developers, regardless of who wrote the initial concept. It is simple, easy to use, and just fun to get their computer work handled by software. Any time a company gets to this great, it is deliberate and requires several revisions to sound and look great is the way it does.

They will be discussing how and why it is important to establish a strong collaboration between designers and developers This is just a quick summary of things that can be done to help teams function together as a better as a whole, but the end aim is to highlight certain key aspects which may be significant for teams.

2.2. A BIT OF BACKGROUND

Simply put, the design-development divide refers to the lack of coordination between designers and developers when the product is being developed. The issues are particularly prevalent in organizations that use strict waterfall methods, in which a planner simply throws the template over the wall and decides it is finished without thoroughly analyzing it because they get lazy with form maintenance.

Without background, it is clarification, developers would have to make up their minds about how to bridge the gap. Because of this, the lack of restriction, the room for off-brand animations, images that go where they should not, even pictures that are not even 50 pixels inside of the corner will be generous. It is known that as long as it is not the bottom line, one has got to size those figures that are deviating from the project hours down to try to see if they add up.

Of course, having design and production teams distant from one another is not just bad for product teams. Even in the absence of initiative from both sides, there are always communication barriers. Experienced designers have their special lingo, and developers have their own words. Furthermore, a piece of new evidence that has been uncovered also supports the anecdotal evidence that shows that a vegetarian is forced to deal with difficulties when trying to order a burrito from a non-vegetarian.

The developers are in charge of the engineering of the product. Any time they discuss, estimate, and introduce a new function, they think about how long it will take, how much money it will cost, and whether or whether they will make it happen. The burden of product, marketing, and distribution managers is usually working to bring new functionality into a product that is in advance of delivery, although the same occurs for developers to hurry up the delivery process.

Concerning product development, most other departments do not know the technological necessities, even though they are supposed to be unaware of them. If the project is almost completed, there is a relentless urgency to send it. The best example of this is the marketing department, which publishes a blog, newsletters, and several other publications and sites on a certain date each week on Monday.

Whenever a calendar date gets off, something has to be readjusted. In addition, Sales has customers who are hoping for new functionality to be provided to buyers, as well as customers who have already committed to the purchase. It is crucial to realize that these pressures are not just imaginary.

Gathering early feedback from customers swift and frequent iterations is always takes precedence for developers as they want to launch their software as soon as possible and work on it out it without regard to their team. To say it another way, piling in more features or extra criteria increases production time, de-depreciates code efficiency, while the liability while shortening test cycles on the other direction of quality.

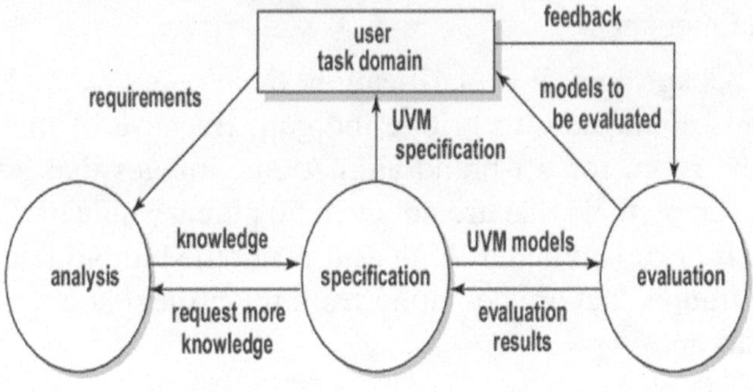

Figure 2.1: Process model of user interface design.

2.3. BRIDGING THE GAP BETWEEN DESIGN AND DEVELOPMENT

One can assume that designers and developers have very different priorities in their minds as they think about what they want to accomplish in their job. After all, it is done, one produces the graphic design and the other is concerned with user experience and restrictions like technologies are essential, but it does not decide what is going to be built. It turns out their intentions should be aligned they all want quality goods.

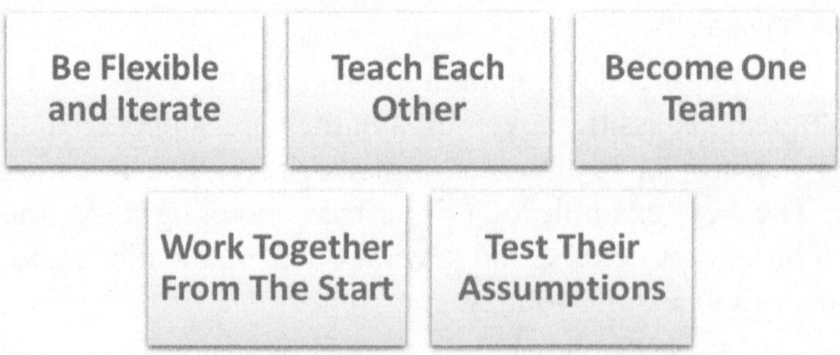

Figure 2.2: Bridging the gap between design and development.

It is not in the interest of mere engineers to improve the things that can be improved; it is of great importance for designers and engineers. To be effective, it is only a matter of using our relevant abilities: programming, planning, technological expertise, and general product proficiency are all ways to arrive at fantastic products that the customers will enjoy.

A selection of methods is provided below to link designers and engineers to make these two groups work together to further expand the use of their skillsets and enhance their collaboration.

2.3.1. Teach Each Other

One must better explain own design decisions to people in full, or they can't comprehend or accept the importance of what one did. They will help the team by taking advantage of this chance. It is only possible for one to learn more if they want to take the time to show them everything they can do. This gives us the freedom to be self-assured, and they do not see themselves as being all one team member, but rather they see themselves as working for specific objectives and ideals.

The resources are the same for both developers and business users and are used in the same way. Teaching their designers sophisticated theories such as universal design goes a long way. This allows planners to think of actual situations instead of only as a question about what might or could happen, not just what could.

2.3.2. Become One Team

This is the team. They want one to repeat after them. This is their team. Leave all the brain theory at the door, and anyone who has managed to make it to this kind of system on the production team will have to do something new. Can this happen to be at the same job as me?

All for one and one and one for all. For me, my coworkers are those with whom I have to spend the most time. They draw on each other for suggestions, then address design elements. They critically analyze wireframes and flows, then search for flaws, then speak about technological concerns, and afterward. They will not be able to make cool things until they are on the same wavelength.

Adding UX is not something one does at the end of production, that is called putting lipstick on a pig. To bring anything to fruition, one must have a strong team of individuals who work together as a team. It is up to us to all of us to do so.

2.3.3. Work Together from the Start

They will put the design at the end of the project. They will bring the design at the end of the project. The following can both be used to explain; They will keep an eye on the developers when they finish the design. There are still some developers somewhere that, if they pay attention, can see where they fail.

Using wireframes and prototypes as the foundation for the design process, they start with using Balsamiq Cloud). They continuously iterate and enlist my product managers and engineers to scrutinize the UI, looking for technological flaws. A job that needs everyone's full attention is on it. They have merged our ideas and capabilities, and abilities, and so the end product is a wireframe that represents all of those factors.

They feel like this job belongs to us, and it represents our best efforts. This is only the beginning: the stage of product development, so be on the lookout for UX specification specifications, several rounds of user and developer roundtables, design evaluations, design prototypes, and so on. Both of them are a part of this process, and they do these things as a team.

2.3.4. Be Flexible and Iterate

The ways one learned as a designer helped save time and provide great results for the client a result has saved me quite a lot of time as a developer. It is acceptable to ship the product and improve on it later. When working with buttons, icons, images, and other similar components, the design files do not need to be pixel-perfect.

Fine-tuning the browser settings or devices to certain specific parameters is a quicker process, which usually produces better performance. This approach requires too several prototypes to figure out how different items in real life are as they are rendered on various platforms, such as Adobe Sketch, Photoshop, etc.

This is also relevant when looking at aspects of the user flow or assessing what is left to be done: features or parts of it that are not yet complete Perhaps, in situations where they have another developer on their side, they can build and code together in tandem. Overall, this results in a closer relationship with their squad, as well as a multitude of suggestions, as well as being quite a lot of fun.

2.3.5. Test Their Assumptions

The whole discussion is meaningless since trying to determine who is right and who is wrong does nothing because this only leads to an infinite cycle of arguments. To begin with, neither side is likely to agree with the other. It does not occur to them that by being extremely critical, do they?

Well, in that case, see it from a positive perspective. To make fine, noteworthy things, they need a great deal of enthusiasm. No groundbreaking products: Lack of passion led to the demise of their products. But, being certain of an assertion, is, to some extent, detrimental because they are not prepared to get proof or do not believe the evidence.

2.4. BRIDGING THE GAP BETWEEN DESIGNERS AND ENGINEERS

Workforce diversification between designers and engineers is a challenge in large organizations where functions can be divided and in silos, where information of each other's activities may be less plentiful. Acting as a full-time developer has shown the problems of both sides of the development equation.

The problems that are associated with working together as a builder have been brought to my attention as a few have been made obvious to me. Just as for all other groups of people, ethnic, racial, or cultural divisions can be erased when they try to relate to one another's position and do our best to see it from another's perspective.

This design document is separated into the areas in which a project can communicate, including what they should tell, the content they should produce, and why they should, and everything they should not do. As well as words that designers may not be familiar with, they have also explained various phrases that serve as generic terms. Lastly, some tips for designers who want engineers to have the best possible experience in working with their construction teams are seen below to improve their ability to feel empathy with their experience.

2.4.1. Say

Once they have developed some useful app states and displays, they must reserve their calendar for when they use it to review user data and expand their app. It is time to speak with the people who will be collaborating with the developers and with other stakeholders. The following items should be considered to support their position:

What: Go over every step of their prototypes and wireframes with the engineers' Product Management and the staff to ensure that everybody is on the same page.

Why: It is helpful to have one-to person one-to-one meetings where they can ask more detailed probing questions and see how the designs can stay true to the initial concepts before implementing the technology in the engineering department. When they are a member of a big organization, it can be difficult to have all the engineers in the same place at the same time to collaborate. First, educate the project manager on their design requirements, and hopes of them being a surrogate for their design team so they are more prepared to support those on their team if the need arises.

What: For the unfamiliar, details (organization structure) must explain any design constraints (such as being bound to one style and format), design decisions (color and font selections), and design specifications (accessibility tooling, localization).

Why: While engineers may struggle to grasp the design decisions, particularly if they were the result of externally-imposed demands, they can count on it if they are operating in environments where ideas are divided up by departmental silos. Empathizing with the engineers and communicating these decisions will minimize the chance that people will make substantive design modifications to their design, so they will better enable the engineers to be better understood and reduce their anxiety overseeing others make alterations.

Write: Using solidly written documents will help drive or prevent the success of their design project. Establish stable, linear timeframes for engineering implementation so that the team can forecast its work accurately and inform the project stakeholders of progress along the way.

What: Describe pixel values and anticipated movements for templates. Create comprehensible layouts, using percentage values and optimistic expectations. Set expectations for layout handling failure.

Why: A product can be written several times to meet the same result in different ways. However slight the variations between these various methods may be, they may add up and allow an overall design to seem contradictory. An expansion that is left incomplete, which might go awry, such as a product design omissions or non-descriptive code, is implemented by the developers who were hired to complete it, such people, which may produce a subpar result.

What: Be sure all staff members have access to the original concept documents in a read-only format. If they have a safer option is to use a central file storage facility to store and transfer the files, and if they do resort to another person to sending their designs over email, always have their designs in the text message thread with them so that they can see them at their fingertips.

Why: It is a problem for big teams to lose track of records, such as emails, as they are sent to only one of the team members at a time (such as a team PM or engineering manager). Storing unmodifiable papers in a central location will reduce the chance of being out of someone being left out of the picture. This could also help the team cut down on the amount of political infighting within the group.

Ensure the designs are not set to read-only, as those who should not be able to modify them can do so by mistake or on purpose.

Want to hear from developers. A commenting mechanism that lets users customize the file instead of making it inflexible is more appropriate. As an alternative, non-designers are encouraged to contribute their ideas and the way they would like to do. It is an example, for example, in the case of Figma, which has a built-in commenting.

Should they be saving their project designs? If they are unsure, Microsoft OneDrive, SharePoint Files, or SharePoint Files, or maybe Google Drive or Teams Embedded Files might be the best options.

What: Configure the paint palette with hex codes for transparency, with various opacities, as well as any adjustments to opacity, and approximate use cases.

Why: This works since all of the settings (such as country, language, currency, calendar, units, et CSS, color scheme, and time zone) are in one place, and they are automatically kept up-to-to-date as well. The details of any project must be well-defined for all interested parties to manage basic changes.

Terms to know:

Hex codes—The use of separate or variable-length coding systems with different colors. A variety of online resources will help is available to them if they choose to create hex codes for their projects.

Variables—Code-reusable values were stored in the function. Both global variables are made available in the software.

2.4.2. Ask

A good designer understands when and how to ask his or her target audience questions. The engineers who will be designing the architecture are, in a sense, the intended customers of the system. Make use of their experience to build more durable product designs.

What: In the app, inquire about popular architecture that can facilitate or obstruct design plans. Is there a tab, for example, that must load in the background before all other pages?

Why: Under the hood, a lovely, basic concept can be extremely complicated. Changing items like the location of a popup or the navigation preferences between pages can be a lot more difficult than they think. At a high level, understanding the underlying architecture will assist them in designing for potential functionality and scalability. Similarly, knowing the fundamental architecture will aid them in communicating their future design goals to the engineers.

What: Inquire about the source of the information. Is it a product of a user's input? Is it from a database or a third-party API?

Why: Designing a static service is easy, but if they start incorporating dynamic data into their experience, they will need to extend their designs to account for issues like inaccurate data sent by customers, network issues when getting data from an API, missing data from a database, and latency issues from external sources. One will properly plan for and design error situations if they have a high-level knowledge of where the application data is sourced.

What: Inside the app, inquire about telemetry. Are there any non-obvious trends that engineers have noticed? Are people, for example, pressing a button at an unusually high rate? Is it common for users to jump from one website to the next?

Why: Though PMs and general user awareness can assist one with their design, telemetry in the app can reveal patterns that are not visible from the outside. Smaller details in their design, such as button location and navigation labels, will benefit from this knowledge. Call out to ask to use one of the equipments at their disposal if such specific telemetry is tucked up in the engineering silo.

Terms to know:

Architecture—The kernel of an application, usually code that is difficult to upgrade because it ensures continuity.

API—Developers use this interface to retrieve information from other services. The Instagram API, for example, helps developers to see how many views a picture gets.

Telemetry—The use and performance logs that an app holds. Typically, data is stored in vast libraries and accessed by personalized dashboards. Anonymization may or may not be possible.

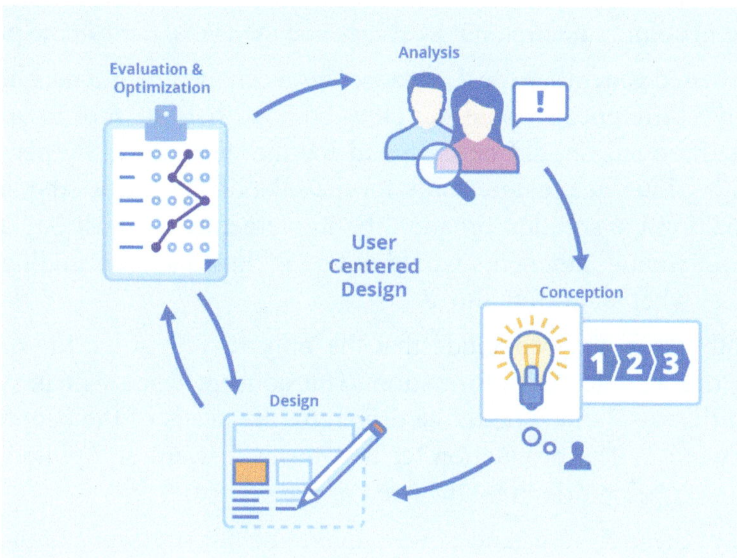

Figure 2.3: User-centered design phases and benefits.

Source: Image by seobility.

2.5. BRIDGING THE GAP BETWEEN DESIGN AND CONSTRUCTION

Design and construction are two separate approaches to the same challenges, but rather than confronting them, and they widen the divide into two entirely distinct worlds. For its architecture, the team must be inclusive, transparent, and helpful, not focused on any one particular group of users; that as many people as possible should be comfortable with the product, and an optimal number of diverse.

The development team wants to see their algorithm straightforward, but delicate handling of algorithm complexity and complexity-expanding aspects of the product to the same customers, to facilitate and streamline their use.

The practice of design and creation involves much experimentation and the construction of concepts. This is simply an endless process of new ideas and discoveries. Doesn't the situation then have to be better? This seems to be the problem that so many businesses have difficulty dealing with: They're interested in communication but unable to see the eye to coordinate projects.

If we had to generalize the differences between their solutions, one might say the two differences would be: How both of their attempt to approach problems. Products should be designed for the needs of customers rather than creating interface features only for the sake of offering aesthetic looks. Before making the solution broader, the implementation must be based on figuring out what components would not be included and on finding a way to get rid of what is limiting those.

In both cases, they conclude that the only way to solve the dilemma is to dig deeper into the information. The concept team will provide an argument that a solution should be driven by the needs of the people using their knowledge. The production team will put forward an opinion on the same solution being driven by the need of the people.

In all of these circumstances, we ought to be both correct and thorough. If a programmer's goal is to make an algorithm and a specific user-oriented application, so he can use the statistical, structured, single-stage approach. But, if the same algorithm is desired for many solutions, designers would need to be exposed to many design stages.

2.5.1. Ways to Get Both Teams on the Same Page

The best approach is not to start with animosity and try and improve but to consider people's strengths and weaknesses so that the tension does not come from those foundations. When programmers are responsible for designing major features, they will choose to go out of their way to add more advanced functionalities that developers are unable to construct themselves. In this manner, the developers find themselves in a bad position.

Likely, developers will not comprehend why something looks good simply because it is technically sound; the reasons it is technically attractive may be irrelevant. Preventing these circumstances is everyone's compromises from being final agreements requires everyone to know his or her limits. More education can be provided to all members of the public and associations by allowing them to share common ideas and capabilities.

Providing the developers with simple descriptions of design features and guidance on how to incorporate them into their work is a great way to improve their understanding of design. The tools for their design team should be able to include an introductory course in programming language syntax and concepts, helping them assess their proposals so-to-and-miles before giving them more complicated tasks. As people acquire more information, they often become more trusting, and fewer misunderstandings are present among the team members.

Many objectives are achieved if the members of the team are clear about what they are striving for and if they can properly express their hopes and fears and frustrations. One way to get designers and developers together is to help developers get started on their design projects and to have designers work with developers as early as possible. But even so, this does not address the issue of developers distracting customers.

Earlyon, have the developers take part in conversations with customers on various aspects of the programs othey can find possible problems before the code is finished and examine the concepts the designers have in advance and get input from developers on all the changes that they would like to make.

Increase the business with all contact for those young designers who would rather send nightly project text updates: Some developers enjoy doing so because it seems as if they can keep their e-mails flowing for a full day or two longer than others believe they can fire them off each time.

As for so many types of communications, some major players might fall out of the radar, and some messages might be overlooked when forms of communication are employed simultaneously. Use a single platform for contact and allow a few spokes to connect that are committed to expanding. To provide another example, they might start sharing e-mail notifications on the first or last day of each week. Make sure designers are connected to developers; afterward, plan the handoff from designers to developers as thoroughly as possible.

To help guarantee that the ease of reuse so they do not have to do all the preliminary work until they get started, specifically arrange the layers according to layer, and group the layers accordingly after the handoff, reach out to the developers to let them know they have got their designs. It empowers developers to examine and qualify the question that they may have, which problems may occur, and the more minor problems to be simplified, making them less likely to remain.

Providing quality assurance support during the development process for new features is like they do for finished products is just as important as developers are doing their work. Bugs in the Webpage and the architecture can easily be found if there are unexpected problems, including the page not matching. This is a very complicated procedure that would take both the creator and the designer's attention. Only a little error such as the misalignment of photos during the test phase of discovery can make it take longer.

Most importantly, learn about other people's history, and use effective resources. Allow individuals and groups to contribute without fear of retaliation. For all material to be readily available, everybody must be able to contact everyone throughout the company. Instead of creating an environment where each developer and designer does their work in addition to each other, utilized an environment where each developer and designer is part of a larger, unified team.

In other words, whether operating on the same function or function is permitted, use representatives of the same department or unit in the same location. It might be nice to get anyone in the workplace to exchange meals and even get them to talk together on occasion. Use special work e-mail or other contact networks for virtual teams. This is useful if they have engineers and designers in various geographical environments, physicalize all project development sessions to help team members from losing track of ideas or discussing problems unnecessarily.

That does not mean that just making the product finish line but instead from idea to code, integrating all of the different stages of the production process in an unbroken manner, right up to and into the handoff. Design and production work in separate halves of the organization, so the term coherent and cohesive product interactions do not make much sense to them. While some gorgeous projects are not readily implemented, a direct contact channel and easy accessibility of knowledge exchange between designers and developers are critical to the launch of popular products.

2.6. THE PROBLEMS IN BRIDGING THE GAP

For sure, the complications that result from a lack of an interpreter are not on their own, however, and are not the only reasons why there is a design production discrepancy. For example, some of the problems most often encountered by people in this group include the following:

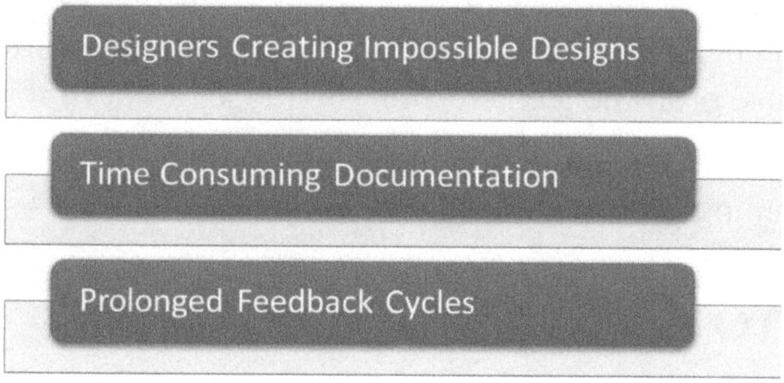

Figure 2.4: The problems in bridging the gap.

2.6.1. Designers Creating Impossible Designs

To someone familiar with HTML and CSS, it is obvious that they are not any different from programmable computer-wrangling whiz kids. Any designers might be drawn to produce ideas in experimental stylish or visually elegant ones that are not transferable, simply because they do not recognize the quirks of CSS and the architecture constraints inherent in the site (easily, or at all).

To close the design-development divide, designers should know enough about CSS to stop developing unworkable solutions.

2.6.2. Time-Consuming Documentation

Using one of the most popular methods to close the design-development gap is to provide metadata: redlines, component diagrams, and many other forms of documentation. Such words are all act as one big deal: The real costs, though, amount to thousands of hours of writing paperwork that only a few people will see or use. Of course, it is not to suggest that they are not useful most software goods benefit from good architecture, including language design as well as design and production documentation.

That said, questions of their usefulness aside, the importance of redlines and other kinds of documents take a long time to produce, and are, as a rule, very tedious. Finding a way to share concept details in less time to prototype.

2.6.3. Prolonged Feedback Cycles

If they plan the app with CSS in mind, then, they must expect user feedback and include thorough and accurate documentation. And no matter what changes there are, it is still beneficial. But when the cycle of repeating actions or iterating processes gets excessively long, it can harm company finances and workers moods Constant input from one period to the next grows, members begin to argue with one another, and therefore the strategies for the group lose prominence, and everyone loses sight of the overall plan.

Bringing the design and implementation into alignment with each other requires eliminating redundant feedback loops.

2.7. WAYS TO BRIDGE THE GAP

Now they understand the problems it creates, let us get down to work. These are the tools they have at their disposal. Look through the list, and they can find the applications to assist them with the design and development tasks they must complete. Even if they are physical talents, certain so-called soft abilities may also contribute. Since there is no way apps are going to fix all of our issues, can they?

In general, though, it is the case that new problem-solving technologies, including the ones that have the advantage of being mechanical, are unable to perform all of the duties that previous ones, such as speaking to people, do not. Thus, with that in mind, three free tools for finding design and construction that do not need spending money are completely ideal for closing the gap between the teams are listed below.

Many contemporary architectures- and application-related organization functionaries are located in distinct structures known as silos. When they were at a larger organization, coordination between the design staff and the production team was almost non-existent. It was as though they were independent properties and worked in separate sections of the house.

At this stage, the designs were final, the designs were handed over to the builders, who used computer technology to integrate the plans without using any methods other than written correspondence. There were no conversations on a day-to-to-day basis; no lateral thinking or networking was going on. This led to 5 main problems:

- This is one explanation for why people found it important to improve the concept several times even though developing the product was pointless.
- The second important thing was that it restricted innovative and inventive development due to the lack of one's domain expertise of the other.
- Three weeks were significantly longer than was predicted, and expanding the QA was an unexpected challenge.
- They had a lot of overlap in skill sets but were unable to use them.
- It is no longer possible to take advantage of new collaboration methods that they did not have previously.

In closing, the following, there is a way to alleviate all of these issues. Is it possible to foster a creative and meaningful design and production collaboration with a singular goal in mind when meeting the requirements of the customer or end-user?

Here are various ways to bridge the gap between design and development.

2.7.1. Communicate Early Often Always

The designers and developers need to be collaborating at all times during a project. Most of all, it ensures they are not forced to deal in a separate Git repository to keep all their ticket data and Sketch designs in sync.

It is not only about conversation but about ideas as well. So, then, designers: Discuss how they are dealing with the new issues with their development team. This means doing the research to find out if the solution is possible. If they have completed their drawings, have them examine places where graphic features cannot be copied for more clarity. determine whether the condition in which real data will destroy the formatting Choose what they believe to be the right names for the architecture layers. Identify potential alternative options with the project team members.

However, make sure to discuss the ways the production team desires for various interactions and interface concepts with them so that they do not get it wrong. They will be able to discuss updates and specifications with their customers until they know the product developers' favorite means of doing so.

Figure 2.5: Simply speaking to colleagues can solve a lot of problems.

Source: Image by Pxhere.

2.7.2. Be Agile

At this time, they are not a cop who regulates processes. So, it is fine if they are not implementing our standard sprint, but do not ensure that they do not ever deviate from it. In my view, there is one element of the flexible approach that any team will still use: building off the last. While it is focused on teams with various subject-matter expertise, it places great emphasis on teams that have specialists around the board.

The process uses design and code collaborations as a foundation for a building. It removes all errors and keeps complications at bay. I think it is a good idea to engage their nice local content manager and the entire cross-functional squad with a day-to-one-run help program at the onset, but that is another story and another.

2.7.3. Speak each other's language

One of the supporters of the "designer should code" philosophy's key points is that it allows programmers to better grasp what their peers do as well as it can better provide them with the knowledge to assist them in figuring out what can be done with the web.

2.7.4. Involve Developers Early in the Process

If they are using Photoshop, freehand sketches, etc., make the design as interactive as possible, yet engage with developers in the design process. Being an excellent site builder is going to allow them to think differently about how stuff could be translated into a web application, leading to a new design perspective. By including them in the design process, they will be able to achieve a smooth integration of design into HTML.

2.7.5. Hand Off by Designers

A critical component of a successful handover from concept to development is included. Often design drawings are developed in Photoshop or some other design application, and those designs are passed to the developers who will be doing the work in HTML, who will then do the necessary conversion work.

(i) Design Files Stay Organized

- When they are finished creating a concept file, mark every single sheet.
- the second option would expand each group to one layer so that it encompasses parts of the page.
- Layer third-party objects or third-party elements inside Photoshop.
- Use the Layer Explorer for creating several sets of nested parts of website pages quickly.
- It allows the creation of layers that can be used to store several times by double-clicking the layer and will open into a new tab.
- Introduce color coding to denote hover states, digital components, as well as expansion and contraction states.

(ii) Walk Through the Designs

When they have finished a draught, have a preliminary meeting with the developers to go through their designs. Conceptually, it can be difficult to distinguish, but it lets them cover minute information that they would otherwise overlook in the process of creation.

(iii) Think Responsive

When planning a graphical asset interface, remember that the app can be seen on a variety of different platforms, small and large screens. Think about how typefaces, relative formatting, and what they will look like on all platforms, such as smartphones, notebooks, and desktops. In this instance, the author has given many assets to go along with the picture. Careful consideration has been given to each product's specific traits, given that they are both concerned with both how long it takes to load and how it appears when it is loaded.

2.7.6. Developer Hand Off

I have seen this in my experience: when the template is in HTML, the results no longer look like what was envisioned. In most cases, such things as an incorrectly formatted font or an icon alignment are not seen until a page is seen for the first time by a person. Until designs are deployed to QA, below are several steps that programmers and designers can take.

(i) Take Full Page Screen Captures

A helpful Google Chrome extension makes it easy to capture screenshots of full screen (it will scroll the full height of the page.) If they are using Google Chrome, then install this extension. Both versions of Firefox, Chrome, Safari, and Internet Explorer can make use of this possible for them in the upper-right corner of their browser window, no matter what.

(ii) Use Skitch

Use Skitch to make comments and provide details for other designers to be added to projects. Often, to highlight concept differences by drawing lines of differing design thickness.

2.7.7. Communication Styles

Communication is vital, followed by tools, then, the economy has an open-minded community that encourages open contact the network has an abundance of low-hanging fruit. Everyone in the company will quickly connect and exchange information with each other, even though it is just little things like messages and documents. To help everyone, here are ideas that have worked in the past and work well for people today's clients.

(i) Be Part of the Daily Scrum with Clients

A large amount of exposure to end-users helped me retain my creativity and think of different ways to appeal to them. Use Slack: I have a Slack app installed on my phone and have a Slack group for my clients to work in inside. Slack is fantastic: It has made it easy to exchange information, has helped me stay connected, and sent me screenshots everywhere I am. It has always been accessible to our developers everywhere.

2.7.8. INCREASE OVERLAP OF KNOWLEDGE

They have found it advantageous to expand my front-end web design skills over time, particularly since they started learning about client-side HTML growth.

(i) For Designers

Have programmers master the coding fundamentals. Have a look here at some resources: Learning to code introduction of web development trying to build a website from this 30-day course, they will learn everything they need to know about HTML and CSS.

In particular, see how different aspects of architecture are manifested in the representation of HTML and all the various frameworks that are accessible through the web.

(ii) For Developers

- Investigate how graphics are used.
- Attending tech meetups and design workshops is an excellent way to expand their horizons.
- Expose the more minute aspects of the argument. As valuable as it is for front-end developers to have an eye for detail, not many possess it. Test the HTML alongside the design. When they are finished, compare the two side by side, try to find the variations.

2.7.9. Create Product Teams versus Functional Teams

Rather than splitting up a team based on purpose, they have seen corporations of several other organizations break up their teams. All in separate locations in the same building: Imagine a team of 15–20 designers, engineers, and QA experts spread out across various locations. Difficult times are on the horizon. Instead, identify those who will be collaborating on a project, and split it down into smaller, more individual teams Any able person should eat.

All of them should stay together if possible. as well as a worldwide team usually have various teams, use different methods of communication including chat applications like Slack, create dedicated networks like Slack for teams to facilitate face-to-to-face interactions.

2.7.10. Share the Latest Design Trends

To keep current as a designer, it is important to remain current on the latest fashion trends. On the subject of design, share insightful stories, mention those that catch their attention, and let everyone else know about the projects they have come across. They are much more influential as a developer if they are aware of recent architecture patterns and methodologies of application implementation than if they are behind them.

2.8. WAYS TO BRIDGING THE GAP BETWEEN DESIGN AND DEVELOPMENT

When it comes to product development, developers and designers have very different responsibilities. Designers are concerned about such features as color schemes and icons on the user interface since the majority of their time is spent on them. Developers take control of all the behind-the-the-the-scenes operations that make a software function often an activity that involves programming work on the developers.

Although designers and developers often do not talk to each other, it is standard to have a high level of incomparability amongst them; it is not to say that the only communication between the two teams is concept files being exchanged is verbal.

This division can occur because employees work on the other two sides of the organization in offices that are on opposite sides of the house, or perhaps, they work in separate buildings depending on where they are performing the tasks that need them to be performed remotely. Regardless of what is dividing designers and developers, creating a more unified connection between them could mean better-designed and built products.

2.8.1. What We Have Got Here Is Failure to Communicate

General targets are easy to miss because they are not defined well. If they are unable to clearly explain them, then varying meanings will lead to further misunderstandings. For this reason, designers and engineers are always coming up with very different ideas when it comes to finding solutions to solve problems.

For example, in the design phase, the designer sets expectations for what the result can look like and behave like and then adjusts these details based on what they wish the customer to get out of tune. The production department may have a different idea about how to bring this picture to life. Design features can make it harder or impossible to finish the product on time. Communication issues can result in numerous problems, including:

- Additionally, there are also times where redundant communication paths between designers and developers.
- Additional and unanticipated testing may even result in loss of revenue, which results in higher losses.
- a perceived failure to use a combination of many existing talents to their full potential.
- If coordination is missed, so more traditional means may be used instead, making it harder for everyone involved.

Let us move on to options. In more detail, below are several things that can be done to close the distance between designers and developers and help improve their contact.

2.8.2. Get Developers Involved Early

in almost all cases, the majority of projects, designers deliver an improved concept document first and then hand it over to the developers. An important way to reduce the distance between designers and developers is to have developers become a part of the process during the early stages of a project. It does not suggest the developers should interfere and become a nuisance.

What they can do, however, is make their developers a part of early steps such as:

- Allowing developers to have a direct conference call with clients.
- accessing insights from mobile application developers to gain knowledge about how a design feature could apply to the web.
- Doing concept exploration and prior checking of the code during the development cycle, rather than after an app is created, ensure any possible or expected problems can be discovered and addressed when they are easier to address.

2.8.3. Orchestrate the Hand Off from Designers to Developers

When it is time for artists and engineers to give their first versions of their computer graphics or sketch designs to other members of the team, a test, set up a procedure to handle it as streamlined and detailed as possible. Through converting such design, web-based designs into HTML, the number of possible problems would be minimized. Make the handoff as planned and productive as possible by:

- Selecting layers and assigning them page sections before them correctly: This assignment must be made before developers get started. It is also true that color-coding interactive components (hover states, click states, etc.) can be as beneficial.
- Explaining the product details to developers: After the handoff, e-mail the designers and ask for more details. This loosens up restrictions on fine specifics and frees developers to inquire about the smallest ones, so they can answer initial questions and work out kinks before the project begins.
- An accommodating design can adapt to whichever screen is used to display the final product or webpage so that users can take advantage of the advantages of the page regardless of what kind of browser they use. A responsive design will involve using different architectures that can be used on all platforms, and applications that allow for efficient use on PCs and mobiles.

2.8.4. Be Just as Prepared with the Developer Hand Off

All things considered, the handoff between the development team and the quality assurance team is standard. The biggest challenge to be addressed is that problems can arise is that the two people working on the project cannot be sure of the same stuff may not be predictable on both sides of the design and development equation. Most tiny problems during QA often cause larger problems later on when objects are misaligned, so it is crucial to ensure that all images are right the first time. Avoid or minimize potential problems with the developer's handoff to QA by:

- Extending full-page screenshot functionality: There are loads of plugins that can be used for obtaining full article text captions. Enabling static analysis at the handoff stage before the project development allows developers to identify possible problems before handover.
- designing rules such as comparing and figure sets of flowcharts with corresponding visuals for valuable supplemental information is beneficial to days in finding flaws and issues with corresponding visual illustrations.

2.8.5. Streamline Communication Efforts

Although this is very common for designers and developers, it is uncommon for designers and developers to use the same communication techniques. Designers who want to communicate with the project regularly but want to send e-mails at the end of the day, might potentially; or they may have designers who want to update project progress through text but want to fire offload it daily.

For instance, where there are several different means of communication being used, certain messages may be skipped and/unread, while other crucial participants may be left out of the loop. Streamline the collaboration processes with their designers and engineers to get them working together with other members of their design and development teams by:

- It can be advantageous to allocate a dedicated channel inbound and outbound Slack for file and screenshot sharing, so they can be passed back and forth freely.
- The inclusion of recommendations for effective contact (e.g., e-mail updates must be shared with everyone at the beginning or end of each workday)
- Using Google Docs, related apps, and attachments such as Google Drive or DropBox to exchange information.
- Conducting frequent meetings with all members of the team to enhance their skills, experience, and network with each other project developments.

Figure 2.6: Streamline communication efforts in the office.

Source: Image by Pixabay.

2.8.6. Boost Knowledge Overlaps

There are misunderstandings and gaps between designers and developers on some of how different processes workflows need one another, causing friction in the workflow process. When designers have new ideas, coding can be hard to use in their designs, so it is possible to implement them, so they do not have coding at all. In other words, since a function or feature can be accomplished, developers do not assume it will be used, particularly if it does not represent the way the user does it in his/sucks.

Help individuals by helping them to master the fundamentals as well as expose them to new ideas and encourage the designers to read about coding may, for example, provide materials on the fundamentals of programming so they can identify if their ideas are feasible to convert into executable code.

Tertiary support for developer assistance can assist the designers in increasing their awareness about design principles and methods, and by providing links to tutorials or other supporting resources they can find online on the subject. Increases in common awareness have the additional benefit of helping to decrease misunderstandings and make contact easier among team members.

Put everyone together who is working on a product instead of having everyone off doing their own thing with their fellow designers or developers. If possible, physically put team members working on the same function together in the same space. Maybe even encourage them to share lunchtimes so everyone will feel comfortable working together. If they have designers and developers in different physical locations, use dedicated communication channels to set up virtual team meetings so concerns can be discussed and ideas can be shared as work progresses.

2.8.7. Exchange Info on Trends

People in the design and development industries are attracted to and stay informed of what is happening in their specific specialization. Ask the designers and developers to make sure to read up on and submit fascinating posts to everyone they know about related to the project to augment their product knowledge. Innovative architecture or deployment techniques may result in a better user interface. Without appropriate creation, construction, the design becomes meaningless.

Without having enough planning in place, the concept or production in place, the design would be severely constrained. Many designers are, by nature, focused on the needs of the customer. Because of this, they are wary of designing a product, which causes them to worry about the end customer. Solution-oriented developers are eager to implement their proposed solution.

It is important to consider the characteristics of the scope of people involved in making their product, especially the weaknesses and expectations of the people using it, to create something useful and attractive for all. They cannot expect that designers and developers will shift their attitudes toward solving problems but what they can do is ensure a continuous dialogue about design and development to keep everyone on the same page, strengthen team members' skills, drive team collaboration, and provide an open environment for sharing new ideas and experiences, and facilitate interaction by being present in the workplace.

2.9. PROVEN METHODS TO BRIDGE THE GAP BETWEEN DESIGN AND DEVELOPMENT

Perhaps the greatest difficulty in web development or digital product growth is bridging the divide between design and development teams. It would be wonderful to have a method that could deal with this challenge. Many good resources are available and can deal with a variety of different problems. It is too early to say which, but it appears that we have not found a tool that deals with all of them.

2.9.1. Understanding the Design Development Gap

In detail, they are concerned about the difference between designers and developers. This issue will likely occur in the course of product growth.

If the waterfall process takes over their business, they will face this challenge. Most of the time, when this happens, it is because the artist believes his job is over until the design is finished. The concept was handed to the artist, who fully intends never to work on it again because he or she has invested too much time and money in making it.

The inventor may dislike the concept because it does not comply with the specification. The only way to ensure this is to get a decent resolution to the problem is to consult with each other first. When the developer does not comprehend the overall purpose of the plan, the project, they can inadvertently communicate something completely different than the developer expected, causing uncertainty.

Furthermore, an earthquake may trigger some failures, such as systems being destabilized. For example, it is an open space for off-brand animations and links that are of poor quality. They might see the corners as being slightly squared, but that does not mean it is. Because of these factors, construction projects and web applications may suffer from both quantity and quality.

And if they believe that just the design/design-development difference afflicts the waterfall organizations, they are mistaken. Additionally, any kind of organization will face difficulties where both the expertise of the veterans and new staff members are missing.

2.10. KEY PROBLEMS PEOPLE MIGHT FACE

There is a slew of problems that arise as a result of the design-development divide. These are a few of the more common issues that their group could face.

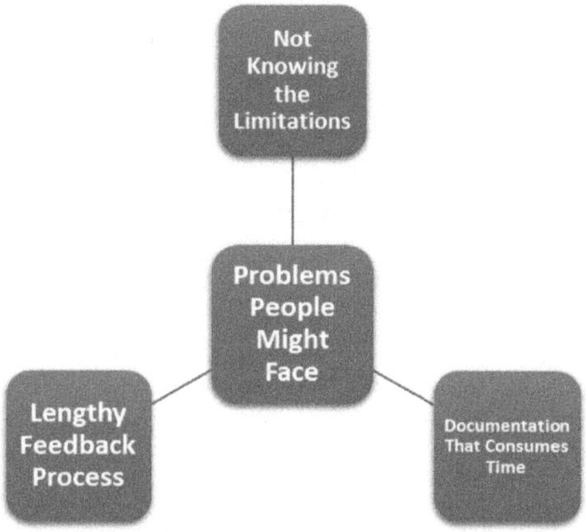

Figure 2.7: Problems face during bridging the gap between in user interface design.

2.10.1. Not Knowing the Limitations

Anyone familiar with CSS is also aware of its shortcomings. It is self-evident that CSS cannot be used to construct anything. Any programmers, on the other hand, are unaware of the restrictions. They have to juggle a lot to get the style they like. They can accomplish this by creating designs in Photoshop or Sketch and assuming that transferring them to the web would be simple. This is almost difficult to do in practice.

Helping the designer grasp the functionality of CSS is one way to bridge the difference here. It is the only way he can stop coming up with a solution that is difficult to implement.

2.10.2. Documentation That Consumes Time

Maintaining adequate paperwork is the most effective way to cross the design-development divide. Redlines, component diagrams, and spec notes are only a few of the documents they will need to keep track of. The records can be referred to by various names, but their intent remains the same.

And if some of these things seem to take a long time, they seem to be time-consuming. Generally speaking, nearly all software goods prosper greatly from the use of architecture, terminology, and implementation metadata.

Since redlines, for example, will take a lot of time and work to build, review and delete any paperwork that is not needed. They should figure out how to document most quickly and straightforwardly possible.

2.10.3. Lengthy Feedback Process

Feedback is a necessary and important component of any project. Even while the designers keep CSS at the forefront of their minds and meticulously record every detail. If the loop is too long, it may harm employee productivity as well as consume the majority of the money. It is not uncommon for them to get all kinds of input from different people. Engineers may lose sight of the overall plan, especially during meetings where stakeholders are at odds. They must figure out how to break the additional feedback loop.

2.11. CONCLUSION

Few references discuss how a designer turns the knowledge learned about users and their jobs into a successful user interface design, even though there are various articles that record aspects of user-centered design. Rather than complex programming technologies, contemporary interactive computing systems have been tightly run for the demands of the customer and the fast launch of a product. Agile techniques and strategies must be combined into the Agile UCD to achieve this trait. The proposed usability-pattern-based requirement-analysis approach has now become the primary expression of rich application software innovations.

REFERENCES

1. Ashwini, A., (2018). *Bridging the Gap Between Design & Development*. [online] Medium. Available at: https://medium.com/swlh/bridging-the-gap-between-design-development-b7f33089a38c (accessed on 6 July 2021).
2. Frontify (n.d). *Bridging the Gap Between Design & Development*. [online] Available at: https://www.frontify.com/en/blog/bridging-the-gap-between-design-and-development/ (accessed on 6 July 2021).
3. Hager, S., (2016). *7 Ways to Bridge the Gap Between Design and Development* | Scalable Path. [online] Scalable Path. Available at: https://www.scalablepath.com/blog/7-ways-to-bridge-the-gap-between-design-and-development/ (accessed on 6 July 2021).
4. Medium, (2017). *8 Tips on Bridging the Gap Between Design & Development*. [online] Available at: https://uxplanet.org/8-tips-on-bridging-the-gap-between-design-development-88c4cf24a0a6 (accessed on 6 July 2021).
5. Medium, (2020). *Bridging the gap between Designers & Engineers—A How-to Guide*. [online] Available at: https://uxdesign.cc/bridging-the-gap-between-designers-engineers-a-how-to-guide-for-designers-616ad6212d21 (accessed on 6 July 2021).
6. Merrill, C., (2018). *Proven Methods to Bridge the Gap Between Design and Development* | Zibtec Blog. [online] Custom Software Development Insights | Zibtek Blog. Available at: https://www.zibtek.com/blog/proven-methods-to-bridge-the-gap-between-design-and-development/ (accessed on 6 July 2021).
7. Williams, J., (2017). *How to Bridge the gap Between Design and Development* | Creative Bloq. [online] Creativebloq.com. Available at: https://www.creativebloq.com/advice/how-to-bridge-the-design-development-gap (accessed on 6 July 2021).
8. Wolfe, A., (2020). *Bridging the Gap Between Designers and Developers*. [online] Goteleport.com. Available at: https://goteleport.com/blog/bridging-the-gap-between-designers-and-developers/ (accessed on 6 July 2021).

CHAPTER 3

USER INTERFACE DESIGN FOR PROGRAMMERS

CONTENTS

In the chapter on user interface design for programmers, several user interface principles have been discussed. It also shed some light on the user interface prototyping. The chapter also discussed usable design for programmers. The chapter also explains the concept of human-computer interaction that is worth considering from the point of view of programmers.

The UI design process is also discussed in detail in this chapter. In addition, evaluation of several interfaces has also been demonstrated to get a better understanding. The importance of user interface tools for programmers is also being discussed in this chapter.

3.1. INTRODUCTION

It is important to note that the recent advancement in the user interface design now considers the entire user experience as part of the design activity. It is not adequate to simply cogitate the user interface, but the entire user experience is the imperative design issue. There are several metaphors of how a user interacts and accesses an automated system.

Seeing these, metaphors back analysts to describe and define the user interface more efficiently. Such metaphors as direct manipulation metaphor, document metaphor, desktop metaphor, and dialog metaphor all define the user interface in somewhat distinctive terms and help in understanding and stipulating an effectual user interface.

It is imperative in the comprehensive design of the user interface to take into consideration good design principles. Such design principles as affordance, visibility, feedback, user shortcuts, error messages, closure, and reversal of actions are all significant no matter what type of user interface is being designed. The judicial use of navigation and menus techniques also makes the user interface more practical.

Often one of the best feasible ways to organize menu items is to group menu items by user or actor role. One powerful technique in the design of the user interface is to make use of storyboards. Storyboards are a set of layouts that mimic the sequence of steps for a specific user process or use case.

Within the past few years, object modeling languages are becoming increasingly complex and expressive. The use of a language such as UML can enhance the potential of user interface designers for modeling interactive systems and application domains.

The recent advancement of techniques that assist in user interface design at the time of object-oriented development process offers programmers with the tools essential to develop vigorous, user-friendly applications. Some of the enhanced techniques comprise preliminary user interface analysis as well as task analysis using interface prototyping that has been proven noteworthy in object identification

It is generally seen that designing user interfaces can be a challenging task. Large development projects will usually hire human interface analysts who are proficient in designing user interfaces.

Smaller projects usually have to depend on programmers to design user interfaces. For the majority of the programmers, user interface design seems to be a frivolous exercise. Although, the user interface plays an eminent role in making customer satisfaction.

A good user interface will be translucent to customers, liberating them to evaluator the system on its output and process. A humble user interface can be an obstruction to an otherwise well-implemented software system. A good user interface accompanies user interface conventions. These may vary between operating systems.

For instance, Microsoft publishes a manual of user interface design guidelines that signifies how users suppose a Windows application to react. Other environments, like the World Wide Web, have their informal conventions. It is generally assumed by the programmers that they understand these conventions, but experience shows that applications do not imitate without a well-established process for user interface design.

In this chapter, the main area of focus will be on designing user interfaces that can accommodate accessibility and ergonomic considerations. Later, one can able to learn how to determine an interface design's potential for success through user testing.

3.2. USER INTERFACE PRINCIPLES

User interface design integrates more than mere aesthetics. A substantial body of empirical data for human-computer interaction has been amassed. This data has been used to articulate many principles at play in contemporary software user interfaces. These principles are the use of self-evident components, metaphors, state visualization, and consistency.

3.2.1. Metaphor

It is always advantageous to design a user interface as a metaphor for a real-world object. The Windows desktop is one of the perfect examples. The desktop has folders that lie on top of it, and folders comprised of documents. If one takes into consideration a metaphor in developing a user interface, be sure to use it steadily across the entire project.

Do not create a metaphor that pertains only to an extremely limited aspect of the program. In addition, it is equally important to realize that while metaphors offer one way to design user interfaces, not all software systems engage metaphors.

Thus, it is always preferable to ensure a good balance between the useless metaphors and arcane and a complete lack of designing interfaces that is pertinent in the real world. As is discussed above, the desktop has become a very efficacious metaphor, and other real-world objects are being employed for programs.

For example, there is work being done on several platforms to have a file system explorer that is a metaphor for a real storage area. Thus, directories having a bigger size on the hard disk will act as larger structures within the program. There are several other metaphors that are used in user interfaces, but it is imperative to retain that not every situation will permit a real-world object metaphor.

3.2.2. Self-Evidence

It is worth noticing that one of the most important features to build into a user interface is self-evidence. The function of several user interface constituents should be readily obvious to even somewhat experienced users. Although, in several cases, the huge number of users will use a system, and each of these users will contact the system from a distinctive viewpoint.

Computer users that are experienced will look for controls that are usually among applications. Many Windows applications have a typical Windows toolbar that embraces buttons to perform common functionalities, such as opening an existing or new document or saving a document.

Experienced users will identify these icons from other applications and feel self-assured to discover the features particular to your application. It is generally seen that beginner users are often frightened of making mistakes. They may not identify these icons.

Their hesitance may deter their potential to use a system that can leave many powerful and useful aspects of a program unexploited. Making the user interface self-evident and customer-friendly eases this fear and accommodates learning.

Figure 3.1 shows the standard Windows toolbar. Some of the icons on the toolbar are instinctive. Most people will comprehend directly that clicking on the image of a printer will cause the present document to be printed.

For example, in the example demonstrated below, two pieces of paper on top of each other may specify "Paste" to experienced users but may mean nobody at all to new users of a system. Although, the Paste Icon is not intimidating and appears to imply that no harm will come to a beginner user for using that facility. Thus, while the accurate meaning of the icon may not be clutched by every user, the information that pressing that icon will not result in posing any significant challenge to a program should be obvious to all users.

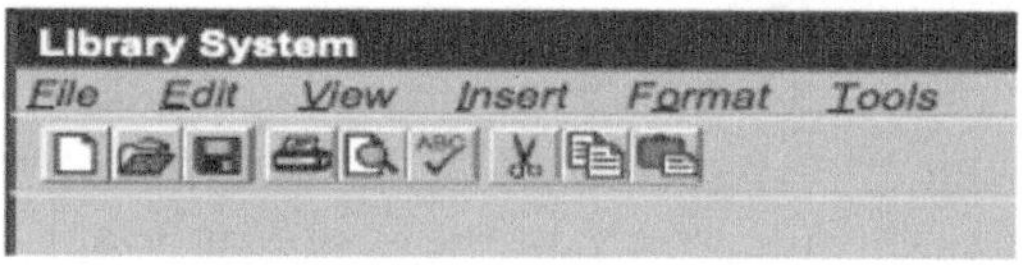

Figure 3.1: Standard Windows toolbar.

Source: Image by cpe.ku.ac.th.

3.2.3. Consistency

It is important to note that the user interfaces should be consistent. There are two types of consistency: external consistency and internal consistency. Internal consistency refers to consistency within the interface of a single software system.

For instance, if the library system permits users to gain access to a patron's account information either by scanning her library card or physically typing her library card number, it must permit users to do the similar when trying out a book to a patron. This kind of consistency makes it easier to learn the system and using the system faster.

External consistency refers to a consistency between a piece of software and the underlying operating environment. For instance, Macintosh applications exposed manifold independent document windows, while Microsoft Windows applications typically execute the multiple-document interface in which a single application window embraces multiple document windows.

Another perfect instance of external consistency between different current versions of Microsoft Windows is the Start Menu. For any Windows user, that user will be aware that programs can be opened through the Start Menu for any version of Windows that has been created since Microsoft Windows 95.

This external consistency in the operating system opens the pathways for smoother transitions for users when they transit between different Microsoft products. A well-designed user interface will follow the agreements of the operating environment in which it exists, making sure to use the external and internal consistencies which are used already.

3.2.4. State Visualization

State visualization refers to the principle that visual cues should be projected to exemplify the change from one state to another. For instance, when a librarian checks a book out to a patron, it is important for him to firstly scan the patron's library card. Following this state, the librarian will either be informed that the patron owes a late fine, or he will begin skimming the book to be checked out.

All of these states should be visually separate so that the librarian can quickly learn to identify each. If the screen informing the librarian of an overdue fine looks too alike to the screen demanding the assets to be checked out, the process will be braked down while the librarian inspects the screen to determine the new state.

A dialog box provides early alert signals to the librarian if the patron owes an overdue fine. The screen that is used to pay an overdue fine is visually distinctive from the screen to be used to enter the assets checked out. These visual representations of the state make it possible for the librarian to work more efficiently and effectively.

State visualization plays an imperative role, but be sure to circumvent producing a modal interface. A modal interface can persist in only one state at a time. For instance, a modal interface might limit a librarian's potential to check out books if he is making use of a system to engage in a time-consuming task, such as generating overdue notices. Modal interfaces are easy to use, but they are rigid.

3.3. HUMAN-COMPUTER INTERACTION

3.3.1. Communication with a Computer

We have to understand our abilities and limitations to be able to design interfaces for humans. Designers need to understand the way communication works between the user as well as a computer. In user interface design, the need for user psychology is crucial.

Humans are not computers, and designers have to understand what the user feels, thinks or what actions they will take to be able to design for humans (Dix et al., 2003, 70–73). Design means to create a non-random effect to solve a problem for the user. It allows users to feel, think and do stuff purposely.

In the year 2014, Marsh stated that understanding all the factors that might affect how a user behaves psychologically and physically is part of user interface design (Marsh, 2014). Human-Computer interaction is communication that takes place between a computer and a human.

It studies computer science and design, user's psychological as well as physiological aspects, the communication between the user and the computer. The user interface is the tool of communication between the computer and the user. Here, the user can mean multiple users, not just a single user on their mobile device or home desktop.

The computer can mean hardware, software, or a combination of that, devices like scales at your grocery shop, ATM, smartphone, or microwave (Dix et al., 2003, 70–73). Our senses help to convey information regarding our surroundings.

The information is sent to human's brains, where it is interpreted as well as built into sensory images. Humans have five major senses; taste, hearing, smell, touch, and sight. A collection of all these senses helps to build our overall view of the world. However, senses of smell and taste do not have a big role in interaction with computers.

Information is primarily received not only by sight but also by ear at the time of using computers. Sounds draw our attention, for example, just hearing the keys while typing or beep of computer while making an error. The role of touch is feeling the phone vibrating or the keys moving.

They are good ways to make the user interface feel more responsive and give user feedback from what they have done. No manual exists that explains the way people operate. Every human being is an individual.

It is hard to design interfaces for humans as they are driven by emotions. Our capability is limited to process information. Understanding the basic human nature and the way humans behave helps to anticipate some of the behaviors and start the design based on them (Dix et al., 2003, 70–73).

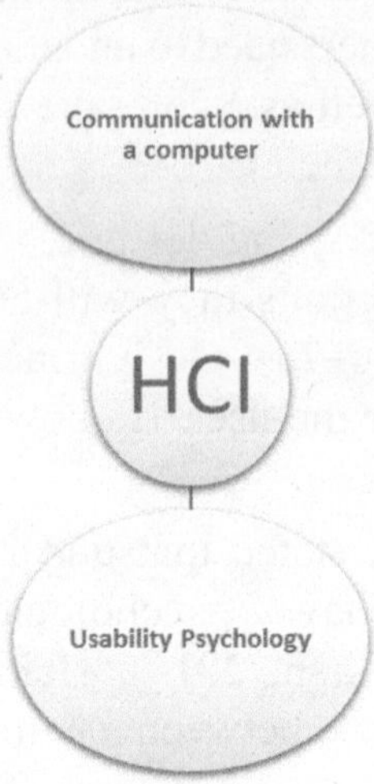

Figure 3.2: Human-computer interaction.

Human beings interact with the outside world through sending information and receiving it with our input as well as output channels. The user sends information to the computer by using hands by hitting the keys, moving a mouse, or tapping the smartphones.

At the time of interacting with the computer, the user receives information that is output by the computer, and the output of the user becomes the input of the computer and vice versa. The senses serve as the main input of humans and motor controls serve as the main output (Dix et al., 2003, 13).

The user operates the computer with the help of the user interface (UI). User interface refers to the communication between the computer as well as the user. The computer is unable to work in the absence of the user.

Computers are developed to make our lives easy in many ways. However, they only work in case they are designed properly. The poorly designed user interface can lead to user frustration, worst-case scenario accidents, and misunderstandings. The reason for misunderstandings might be that the communication is not clear or the user does not understand the interface.

The user interface can be both software as well as hardware. Hardware refers to all the physical parts of a computer and related devices, for example, the mouse. The communication between a computer and a human being should feel as fluent as talking to another person, in case computers are made for humans.

A human might communicate to computers without even noticing it. Nowadays, even cars can have software and as a result, they run on code. A user does not pay any attention to the user interface design, except in a situation where they bump into an error. As it feels natural to use, it is harder to spot good design (Norman, 1988; Saariluoma, 2004, 15; Dix et al., 2003, 3–4).

The 1st generation of iPhone was released in year 2007. Before that time, all the smartphones had hardware keys.

Apple was the first to remove most of the hardware and replace the same with a bigger screen. With the introduction of new design as well as technology, it became possible to replace the hardware with software. In the year 2007, Jobs stated that software is easy to update so that lifespan of product can be longer and design possibilities are unlimited.

Designing for humans includes prototyping, the cycle of designing as well as evaluation of the interface. Prototyping can be just wireframe paper models, and it is fast, cheap as well as a crucial part of user interface design. It is necessary to prepare the design in theory before developers start building the software. It is faster as well as cheaper to test the product, but it is still a paper model. One of the important rules is to fail fast and learn from it.

Wireframe models are very easy to test immediately and to change if needed. Prototyping is the simple way to understand your users together with their requirements. Sometimes it is very difficult to put yourself in the position of users. The interface is very familiar for designers but not for the user who sees it for the first time. Everything makes sense to the designers because it is built by them and have been working on it for a long time.

3.3.2. Usability Psychology

Human beings give reactions with feelings as well as emotions. Moods have a great impact on the way we behave. Our moods and emotions keep changing all the time, and they have an impact on our thinking, perceiving, problem-solving, the way we memorize things, and how effective we are. Emotions can be classified into two categories that are positive as well as negative.

All the feelings that have a positive impact fall under the positive category and all the feelings that have any negative impact fall under the negative category.

We can operate better if we feel positive. Therefore, we can say that positive feelings have a positive effect on human beings. When humans feel good, they become more creative and find better solutions for problems.

Human beings do not think only at this moment, and here, we think in time. We can anticipate as well as plan the future and remember the past. Our experiences of the past influence the way we see the future and live in the present moment.

Our mood is affected by time. In case we expect or want something but we are unable to get it, that will make us feel frustrated or maybe even mad towards whatever is stopping us. Feelings are of two types: motivation as well as emotion. The term motivation denotes what we want, and the term emotion denotes the way we feel. (Sinkkonen et al. 2006, 250–256; Marsh, 2014.) To use the product, the user needs to have a goal.

The goal should be clear, simple, and attainable, like reading the e-mail. The user is required to be motivated to achieve this goal. Motivations are physical or psychological needs. The user can be aware or unaware of being motivated. Therefore, the experience of motivation can be conscious or subconscious. Motivation is easy to gain but also easy to lose.

Designers should try to keep users motivated in a user interface (Sinkkonen et al., 2006, 263–268; Marsh, 2014). Humans need memory to process information. Memory plays a vital role in making plans and further developing them. A memory of humans can be classified into three parts and they are sensory registers, short-term memory as well as long-term memory.

Sensory memory can register all the information, however, forgets it in a couple of seconds. Short-term memory is also known as working memory. In this type of memory, the information can be stored for up to 30 seconds, and afterward, the information is either forgotten or stored in the long-term memory (for days to years). The information can be forgotten or disappear at any stage. To be able to function effectively, forgetting information is very important.

Forgetting is considered as a negative thing. However, humans would live in information chaos in case, they could store all information. It would be very hard to categorize information, connect it as well as learn. One of the most important parts of remembering is forgetting.

Memory helps us to learn things. Information needs to be processed and understood in the working memory so that it can be stored in the long-term memory, and therefore, it can be used later if necessary.

Remembering can be made easier by repeating the information and connecting it to already learned information. By putting information in categories, we can connect it and understand it in a better way. Connecting as well as categorizing information is known as schema. A schema is a cognitive framework or concept that helps organize and interpret information.

The schema does not hold the subject's detail but it is more like concept descriptions. It also holds supposition of the subject together with the relationship with various other subjects. The schema is wireframing information in the minds of human beings.

The individual provides the basis for the content of the schema (Sinkkonen et al., 2006, 167–170). Schemas can contribute to stereotypes and make it difficult to retain new information that does not conform to our established ideas about the world.

Learning refers to storing information in the memory. It is a process where the learner forms an idea or a mental model of skill and can apply as well as use it to a new situation or practice the skill as long as they can repeat the action purely.

In 2006, Sinkkonen stated that learning for humans is a feature that does not disappear at any point of active life. A user has to perceive all the elements in the user interface that are essential for the task to use the product.

The users have to be able to follow the way their actions influence the user interface. In case the user does not notice the necessary parts to complete the task in the interface, either something else is taking the attention of the user or the user perceives it in a wrong way.

Perceiving does not include just sensing as it is not enough that all the essential parts are in the user interface; however the user must be able to perceive them and recognize them as something before, they can utilize the information.

3.4. USER INTERFACE PROTOTYPING

A prototype can be a simulation of the actual system. User interface prototypes are taken into usage to test the usability of the interface. A prototype is a tool for expressing and refining necessities.

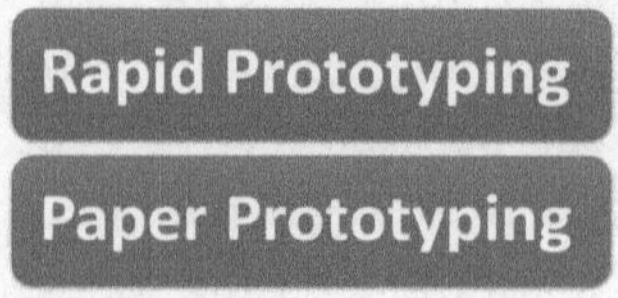

Figure 3.3: Types of user interface prototyping.

A prototype may take the usage of a piece of software that implements the interface and simplest functionality of a system, or it may be as modest as a set of screenshots known as a paper prototype. several classes of prototypes are discussed as follows:

- **Horizontal prototype:** implements a wide range of features and a shallow application of each feature.
- **Vertical prototype:** It implements a subgroup of the features with more thorough implementation.
- **High-fidelity prototype**: It is a software prototype designed to gaze and function similarly as the final software system.
- **Low-fidelity prototype:** a non-software prototype that uses documents, diagrams, or presentations to commune the user interface. Each of these programs of prototypes can be formed by either using the paper or rapid prototype process. The two prototype development processes are discussed below.

3.4.1. Rapid Prototyping

Rapid prototyping can be defined as a process by which a working demonstration version of a piece of software is produced in a very limited time. When designing a user interface, several rapid software prototypes may be urbanized and verified with users before a final user interface is decided upon. A software prototype is never formed into a fully operational system.

The prototype can be used only for the demo objective. It communes a vision of the final product's entrance, and it aids as a catalyst for new ideas. Software prototypes are generally formed developed using Rapid Application Development (RAD) techniques.

No emphasis can be placed on performance. Software prototypes can be developed with the help of using Rapid Application Development tools such as Visual Basic or Delphi even if the actual software system will be industrialized by using another language.

3.4.2. Paper Prototyping

Paper prototyping can be taken into use to communicate similar information as a software prototype without the cost of developing a working piece of software. Paper prototypes use storyboards and screenshots to exemplify the system's functionality and appearance. Paper prototypes may be part of a formal interface requirements document.

3.5. USABLE DESIGN

An effective usable design comprises three distinctive designs that are imperative for each aspect. They are:

- Interface design;
- Information design;
- Interaction design;

3.5.1. Interface Design

To attain a perfect design all three designing characteristics should be comprised and overlapped. Failure of any one of these designs will result in the failure of the whole design. A system with essential information and interface design will not be efficacious unless there is great communication with the user.

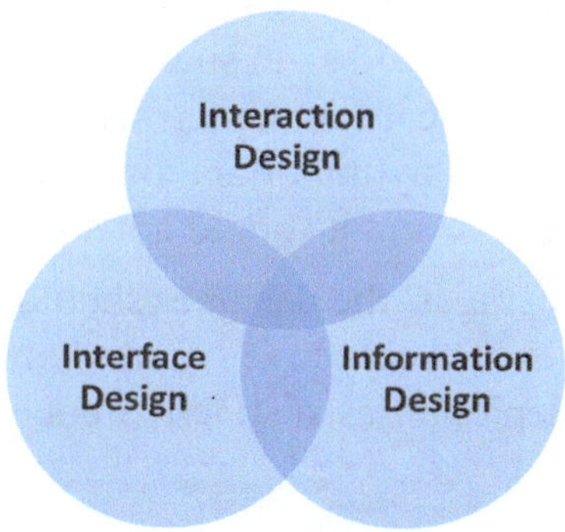

Figure 3.4: Various types of usable design.

It is worth noticing that the interface should be designed with adequate objects and colors. Colors should be used restrictively with a meaning. There is a need to take into consideration Proper color pairing, and at the same time, people having color blindness should be taken into account. Objects such as menus, buttons, and icons should be used.

Navigation can be made easy. A proper style of interaction should also be followed with suitable symbols and language. A bad interface will result in the production of a useless system. 'Golden' rules to be followed in creating an effective interface design.

3.5.2. Information Design

Present the information appropriately so that it would be easy for the user to recognize who is the intended recipient, an adaptation of information, graphical objects, text, diagrams as well as photos. Information for the relationship between values and change of information should also be existent.

The use of colors should also be inadequate way when showing the transition in the system status, and color-coding can be used to provide a way for users to identify the task they are executing. Colors should be used in a controlled manner when it comes to dark colors, and they should be used creatively and thoughtfully.

Proper emphasis should also be given to roper color pairings. The system should be able to foresee consistently what the user is doing and give messages pertinent to the situation, experienced users would not prefer long messages and new users might not be able to get sufficient information from short sentences, thus so both types of users should be taken into account and let them use message conciseness.

The layout of messages should be in such a manner that different classes of users would be able to handle it contingent on their skills. While expressing thought in a message, it is highly important to tailor it so that it gives out the message in an optimistic way and it should not sound humorous.

When giving out messages, the designer should be cognizant about the culture being followed in the respective country where the system is being sold to give a proper depiction as distinctive countries follow distinctive cultures.

3.5.3. Interaction Design

When designing an interface system, several things should be taken into consideration, such as the age, background, and culture of the user, the user's computer knowledge should be evaluated, and also his potential to adapt to the system.

Navigating and finding help should be made easy for the user while looking for any information. Help should be offered for the function, to interact and learn. The use of distinctive sounds, lighting, can be done to fascinate the user. The user should not feel any pressure or stress.

3.6. THE UI DESIGN PROCESS

The UI design is a process where the users interact with the designers. There are three imperative needed activities in this design process.

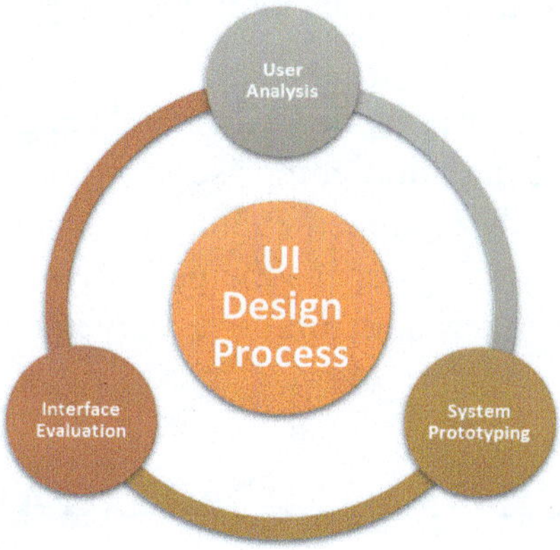

Figure 3.5: The user interface design process.

3.6.1. User Analysis

In this analysis process, a task has been developed that the users do in other words comprehending what the users do with the system. To develop an effective interface, we have to cognize what the users want to do with the system. User analysis should be depicted in such a way that it is simply comprehended by the users and other designers. Scenarios are one of the ways of defining the examination.

Some of the necessities from the scenario are helped in using adequate search terms as the users may be unconscious of them. Searches and requests for copies of the allied material must be carried out by the users.

3.6.2. System Prototyping

The development of a prototype system plays a very important role in guiding the evolution of the interface. The prototyping process with early paper prototypes is used as the foundation for automated prototypes.

3.6.3. Interface Evaluation

In interface evaluation, the users' experience with the interface has been amassed. Evaluation of interface design should be done to know its aptness. Full-scale evaluation is not applicable for the systems and is also extremely costly. Usually, the interface is assessed against the usability specification. Interface evaluation contingent on usability attributes evolution is conducted.

Assessment of user interface design in a systematic way could be a high-priced process. This process comprises cognitive scientists and graphics designers. A cheap way to assess an interface is by reviewing users via a questionnaire.

Users should rate the questionnaire with their background and experience. The observation-based evaluation comprises of users watching as they are using the system. Specially equipped evaluation is needed for whole video analysis. Analysis of recordings opens the pathways for designers to find if the interface needs too much unnatural eye movement or hand movement will do.

3.7. IMPORTANCE OF USER INTERFACE TOOLS FOR PROGRAMMERS

It is generally seen that there are several advantages of using user interface software tools. These can be categorized into two main categories. First, the quality of the resulting user interfaces should be higher, for the given motives:

Designs can be rapidly prototyped and executed, probably even before the application code is written. This, in turn, resulted in more quick prototyping and thus more repetitions of iterative design, which is a critical constituent of attaining high-quality user interfaces (Nielsen, 1993b). The consistency of the user interface will be advanced because the code for the user interface is formed mechanically from a higher-level requirement.

Distinctive applications are more likely to have reliable user interfaces if they are formed using the same UI tool. It will be calmer for an array of specialists to be comprised in planning the user interface, instead of having the user interface formed completely by programmers.

Graphic artists, usability specialists, and cognitive psychologists may all be comprised. In specific, expert user interface designers, who may not be programmers, can be the responsibility of the overall design. More exertion can be used on the tool than may be practical on any single user interface, as the tool will be used with several distinctive applications.

Help, Undo, and other features are more probable to be obtainable because the tools might support them. Second, the UI code might be simpler and more reasonable to form and maintain. It is primarily due to the following: Interface specifications can be validated, represented, and evaluated more easily. There will be less code to write, as the majority is conveyed by the tools.

There will be enhanced modularization, because of the separation of the UI component from the application. This should permit the user interface to alter without impacting the application, and a large class of changes to the application (such as transitioning the internal algorithms) should be conceivable without affecting the user interface.

The level of programming proficiency of the interface implementers and designers can be lower, as the tools conceal much of the intricacy of the underlying system. It will be simpler to port an application to distinctive software and hardware environments because the device dependencies are remote in the UI tool.

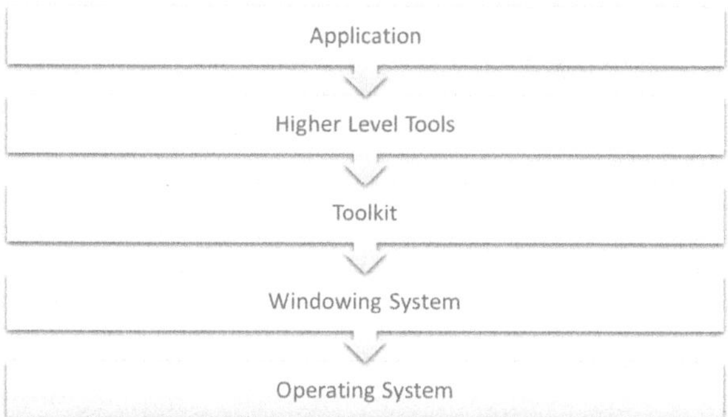

Figure 3.6: The components of user interface software.

(i) Overview of User Interface Software Tools

As it is well known that user interface software is so challenging to create, it is not astonishing that people have been functioning for a long time to produce tools to help with it.

Today, a variety of these ideas and tools have advanced from research into commercial systems, and their efficiency has been sufficiently confirmed.

Research systems are also likely to evolve rapidly, and the models that were widespread five years ago have been made outdated by more effective tools, variations in the computer market, and the occurrence of new styles of user interfaces, such as multimedia and handheld computing.

3.7.1. Components of User Interface Software

UI software may be categorized into various layers: the toolkit, the windowing system, and higher-level tools. Of course, several practical systems span multiple layers. The windowing system backs the split of the screen into distinctive (typically rectangular) regions, known as windows.

The X system (Scheifler, 1986) categorized window functionality into two layers: the window system, which is the programming or functional interface, and the window manager, which is the user interface.

Thus, the window system offers procedures that permit the application to frame pictures on the screen and get input from the user; the window manager lets the end-user transfer windows around and is accountable for exhibiting the title lines, icons, and borders, around the windows.

Although, several people and systems use the name "window manager" to suggest to both layers, as systems such as Microsoft Windows and Macintosh do not distinguish them. It uses X terminology, and uses the term windowing system to denote both layers.

It is important to note that Microsoft bewilderingly calls its whole system Windows (for instance, Windows XP or Windows 98). This comprises several distinctive functions that here are distinguished into the operating system part (which backs file access, memory management, networking, etc.), higher-level tools, and the windowing system.

On top of the windowing system is the toolkit that comprises several normally used widgets (also known as controls) such as buttons, menus, text input fields, and scroll bars. On top of the toolkit might be higher-level tools that provide the opportunity to the designer to use the toolkit widgets. In the next following sections, each of these components is discussed in more detail.

3.7.2. Windowing Systems

A windowing system can be defined as a software package that opens the pathways for users to control and monitor diverse contexts by dividing them physically onto distinctive parts of one or more display screens (Myers, 1988b).

Although a majority of current systems offer toolkits on top of the windowing systems, as will be elucidated later, toolkits normally only address the drawing of widgets such as menus, buttons, and scroll bars. Thus, when the programmer is interested in framing application-specific parts of the interface and lets the user influence these, the window system interface can be used straightforwardly.

Therefore, the windowing system's programming interface has an important impact on the majority of user interface programmers. The first windowing systems were applied as part of a single system or program.

For instance, the Smalltalk (Tesler, 1981), the EMACs text editor (Stallman, 1979), and DLISP (Teitelman, 1979) programming environments had their windowing systems. Later systems applied the windowing system as an essential part of the operating system, such as SunView for Suns, Sapphire for PERQs (Myers, 1984), and the Macintosh and Microsoft Windows systems.

In a way to permit distinctive windowing systems to function on the same operating system, some windowing systems, such as Sun's NeWS and X (Gosling, 1986), operate as a distinct process and use the operating system's inter-process communication method to join to application programs.

3.7.3. Structure of Windowing Systems

A windowing system can be cogently categorized into two layers, each of which has two distinctive parts. The window system, or base layer, applies the fundamental functionality of the windowing system.

The two portions of this layer manage the display of graphics in windows (the output model) and the access to the several input devices (the input model), which normally comprises a pointing device and a keyboard such as a mouse. The main interface of the base layer is procedural and is known as the windowing system's application programmer interface (API).

The other layer of the windowing system is the user interface or window manager. This comprises all aspects that are noticeable to the user. The two slices of the user interface layer are the presentation that embraces the pictures that the window manager exhibits, and the commands, which are how the user influences the windows and their contents.

3.7.4. Base Layer

The base layer can be defined as a procedural interface to the windowing system. In the 1970s and early 1980s, there was an extensive number of distinctive windowing systems, each with a distinctive procedural interface (at least one for each hardware platform).

People who are writing software found this to be intolerable as they wanted to be able to run their software on distinctive platforms, but they would have to rewrite huge sums of code to convert from one window system to another.

The X windowing system (Scheifler, 1986) was formed to resolve this problem by offering a hardware-independent interface to windowing. X has been fairly efficacious at this, and it pushed all other windowing systems out of the workplace hardware market.

X will likely be popular as the windowing system for Linux and all other UNIX implementations. In the rest of the computer market, the majority of the machines use some form of Microsoft Windows, with the Apple Macintosh computers having their specific windowing system.

3.7.5. Output Model

The output model can be defined as a set of procedures that an application can use to draw pictures on the screen. All output must be guided through the window system so that the graphics primitives can be trimmed to the window's borders.

For instance, if a program frames a line that would lengthen beyond a window's borders, it must be trimmed to make sure that the contents of other, independent, windows are not overwritten. The majority of the computers offer graphics hardware that is enhanced to work proficiently with the window system.

In early windowing systems, such as Sapphire (Myers, 1986), and Smalltalk (Tesler, 1981), the main output operation was BitBlt (also known as RasterOp, and now sometimes CopyRectangle or CopyArea). These early systems mainly supported monochrome screens (each pixel is either white or black).

BitBlt takes a rectangle of pixels from one chunk of the screen and replicas it to another part. Several Boolean operations can be stipulated for uniting the pixel values of the source and destination rectangles.

For instance, the source rectangle can purely substitute the destination, or it may be XORed with the destination. BitBlt can be used to draw solid rectangles in either white or black, scroll windows, display text, and perform many other effects (Tesler, 1981).

The only added drawing operation usually maintained by these early systems was sketched straight lines. Later windowing systems, like the Macintosh and X, added a complete set of drawing operations, such as unfilled and filled polygons, lines, text, arcs, etc.

These cannot be executed using the BitBlt operator. With the growing reputation of nonrectangular primitives and color screens (such as rounded rectangles), the use of BitBlt has been decreased significantly. Now, it is chiefly used for copying and scrolling off-screen pictures onto the screen (e.g., to apply double-buffering).

A few windowing systems provide the way for a full PostScript imaging model (Adobe Systems Inc., 1985) to be used to produce images on the screen. PostScript offers device-independent arbitrary rotations and coordinate systems and scaling for all objects, comprising text.

Another main benefit of using PostScript for the screen is that a similar language can be taken into use to print the windows on paper (since several printers admit PostScript). Sun created a form used in the NeWS windowing system, and then Adobe (the maker of PostScript) came out with an approved version known as Display PostScript that can be used in the NeXT windowing system.

An alike imaging model is offered by Java 2D (Sun Microsystems, 2002), which is applicable on top of (and hides) the underlying windowing system's output model. All of the standard output models only comprise drawing operations for 2-D objects. Extensions to support 3-D objects such as OpenGL, PEX, and Direct3-D. PEX (Gaskins, 1992) is an extension to the X windowing system that integrates the majority of the PHIGS graphics standard.

OpenGL (Silicon Graphics Inc., 1993) is centered on the GL programming interface that has been taken into use for several years on Silicon Graphics machines. OpenGL offers some machine independence for 3-D as it is offered for several X and Windows platforms.

Microsoft conveys its own 3-D graphics model, known by the name Direct3-D, as part of Windows. The earlier windowing systems presumed that a graphics package would be executed using the windowing system.

For instance, the CORE graphics package was applied on top of the SunView windowing system. In addition, systems like the Macintosh, NeWS, X, NeXT, and Microsoft Windows applied a high-tech graphics system as part of the windowing system. Now, with Java3-D and Java2D, in addition to the Web-based graphics systems like the VRML for three-dimensional programming on the Web (Web3-D Consortium, 1997), one can expect a return to a model, with the graphics on top of the windowing system.

3.8. EVALUATION OF USER INTERFACES

In this section, the examples of different User Interfaces will be taken and assessing them based on the quality measures pondered in the previous section.

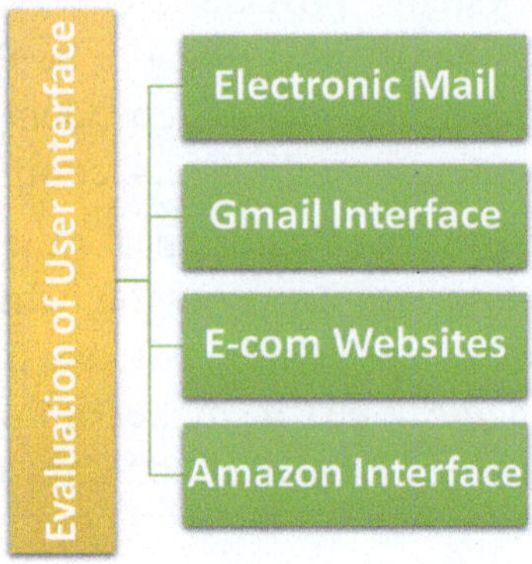

Figure 3.7: Evaluation of user interface.

3.8.1. Electronic Mail

It is generally seen that electronic mail (e-mail) systems are the most widely used interfaces of our time. They are almost free, in addition, premium services are available at an economical rate. It offers effective online communication. It is worth noticing that there is a wide range of E-mail Service Providers (ESPs). It is worth studying three of such ESPs and then assessing the UIs presented in them.

3.8.2. Gmail Interface

First of all, we take into consideration the Google Mail (Gmail) interface. It is possible to move from one measure to another in a way to assess whether it conforms with the quality characteristics or not.

- It is generally seen that the Gmail homepage has the persistent display of the status of the byte increase in the storage space being offered to the users. This offers the user to view the system status graph for the continuous improvements in the offered user space.
- All the UIs of Gmail are intended under the same choice of reliable color schemes, font style, font size, etc. Henceforth, there are no blows to the user.
- During the operation of Gmail, there are several possible scenarios in which there are chances that a varied error can occur. These errors comprise sign-in issues, delay in response, server conflict, etc. In such a case, Gmail offers efficient and fast error handling. For instance, if we take one specific scenario, in which while there is an interruption in getting a response from the server for sending an e-mail because of the slow speed of the internet connection. The UI updates its status regularly. This specific feature allows the user to know that the system is working on the request, but, because of the sluggish transfer of data, is unable to serve the request straightaway.
- It is generally seen that Gmail interfaces are rapidly loaded on browsers because they are intended in such a manner that they are not great in size. Consequently, they offer high speed in loading. Another unique and attractive feature that Gmail has is the provision of a much lighter-weight Hypertext Mark-up Language (HTML) view.
- By making use of this view, the users can look for information even if they are using a very sluggish internet connection such as the traditional dial-up connections on 56K modems. The HTML view is much like the unique view and is much faster and lighter.
- The Gmail interface exhibits simple menu options that are disseminated in the column located at the left-hand side of the inbox panel. Thus, locating the needed options on Gmail is comparatively easy and modest for a user.

- Gmail provides comprehensive help options for all its features. The help is a step–by–step guide for the solution to the problems that are expected to be faced by the user.
- It is generally seen that the Gmail interface uses less color and more speed and quality. Although everything is not white and black, colors are suitably employed in such a manner that they are frustrating for the eyes of a user.
- A user can find the pertinent and noteworthy amount of information at pertinent areas of Gmail UIs. The linked information is spread across the concerned sections of pages which makes it tranquil to extract and track.
- Last but not least, the Gmail UIs exemplifies user comfort and ease in addition to simple and fast access to the target utility.

3.8.3. E-Commerce Websites

- In addition to the type of websites discussed above, other websites are used for online shopping, transactions, banking, etc. In such websites, the UI plays an eminent role in the success and further progress. Overall, there are two sorts of such websites:
- The websites that are signifying a business are already existent in the real market offline, with the online business serving as an extension to the business.
- The websites that are signifying the business which only exists offline, with only the delivery and shipping system, etc. present office.
- In the latter, adequate UIs are the most critical part of the business. They serve as the critical object for enticing customers and growing business. In the following sections, the major area of focus will be e-commerce websites.

3.8.4. Amazon Interface

- The UIs of Amazon are intended in such a manner that the customers are invigorated to purchase goods and feel happy with the accessible interface utilities and design.
- There are distinctive sorts of items available at Amazon for purchase. These items are positioned at adequate places in

distinctive sets so that it is cool for the user to position and buy them.

- Searching techniques for rapid item recovery are offered, on the top of the homepage UI. In this manner, the user sees that the core stress is given on the penetrating of the item. A user retrieving the information on an item for buy using this utility.
- The entire website uses a similar template and looks for displaying pertinent information. In this manner, nothing seems to be out of perspective concerning the outlook.
- The chief focus is given to the products. Thus, the websites are extremely cool to load and fast for the processing that is needed.
- The fast query response time permits the user to signify and provide a query for buying in no time.
- The website is intended by taking into consideration that the presentation of the UI should not be annoying to the user. Therefore, no sharp contrasts can be used, but, instead, the simple and plain background can be used so that the main area of focus will remain on the product only.

3.9. CONCLUSION

In the end, it is concluded that there are several tools for user interface design that should be taken into consideration while designing UI design by programmers. The needs and wants of consumers should always be in the mind of programmers while designing a user interface. There is a proper design process for the user interface that is worth considering.

In addition, it is always recommendable to engage in the prototyping of the user interface to address any shortcomings, if there are any. There are also some principles of user interface that help in making the well-designed user interface such as metaphor, self-evidence, consistency, and state visualization.

REFERENCES

1. Chandra, R., & Guntupalli, C., (2021). *User Interface Design- Methods and Designs of a Good User Interface Design*. [e-Book] Available at: http://www.diva-portal.org/smash/get/diva2:215020/fulltext01 (accessed on 6 July 2021).
2. Cybulski, P., & Horbiński, T., (2020). User experience in using graphical user interfaces of web maps.*ISPRS International Journal of Geo-Information, 9*(7), 412.
3. *Designing the User Interface*: *Chapter 8* (2021).[e-Book] Cengage Learning. Available at: https://www.cerritos.edu/dwhitney/SitePages/CIS201/Lectures/IM-7ed-Chapter08.pdf (accessed on 6 July 2021).
4. *Human Factors; User Interface Design Principles for the Telecommunications Management Network (TMN) Applicable to the 'G' Interface*. (2021).[e-Book] Available at: https://www.etsi.org/deliver/etsi_eg/201000_201099/201024/01.01.01_50/eg_201024v010101m.pdf (accessed on 6 July 2021).
5. *Interface Design and Prototyping*. (2021).[e-Book] Available at: https://www.cpe.ku.ac.th/~plw/oop/e_book/ood_with_java_c++_and_uml/ch9.pdf (accessed on 6 July 2021).
6. Joel on Software, (2004). *User Interface Design for Programmers*. [online] Available at: https://www.joelonsoftware.com/2001/10/24/user-interface-design-for-programmers/ (accessed on 6 July 2021).
7. Mishra, P., (2013). *User Interface Design*. [e-Book] Helsinki Metropolia University of Applied Sciences. Available at: https://core.ac.uk/download/pdf/38089589.pdf (accessed on 6 July 2021).
8. Myers, B., (2021). *Graphical User Interface Programming*. [e-Book] Available at: https://www.cs.cmu.edu/~bam/pub/uimsCRCrevised_C360X_CH48.pdf (accessed on 6 July 2021).
9. Pela, P., (2021). *Introduction to Coding for UX and UI Designers: Data Structures and Objects*. [online] Medium. Available at: https://uxplanet.org/introduction-to-coding-for-ux-and-ui-designers-data-structures-and-objects-88eb219366c5 (accessed on 6 July 2021).
10. *User Interface Programming*. (2009).[e-Book] The IDL®, IDL Advanced Math and Stats™. Available at: http://www.geo.mtu.edu/geoschem/docs/IDL_Manuals/widgets.pdf (accessed on 6 July 2021).

CHAPTER 4

VISUAL INTERFACE DESIGN

CONTENTS

This chapter explains the basic concept and ideology behind visual interface design. This chapter also explains the importance of visual design in aesthetics and strategic design. The chapter on visual interface design provides highlights on the visual elements that play a significant role in UI design. This chapter also addresses the various fundamental elements of an interface design. This chapter sheds light on the various types of user interface design tips that play an essential role in designing. This chapter also mentions the different principles of visual design in user experience.

4.1. INTRODUCTION

Visual design targets to enhance the design's/product's aesthetic appeal and usability with fitting pictures, typography, space layout, and color. Visual design is more than attractive. The elements are carefully put by designers to make interfaces that drive conversion and advance the experience of users.

4.2. VISUAL DESIGN IN AESTHETIC, STRATEGIC DESIGN

Users take typically 50 milliseconds to form first impressions. The fact shows the visceral state of emotional design- in other words, the first impression or a gut reaction of a user to a design. They will leave if they don't like what is shown to them in those critical moments. So, as a designer, visual design is an important part of work. Many people misinterpret to mean those attractive elements are included by designers to increase an item's attractiveness.

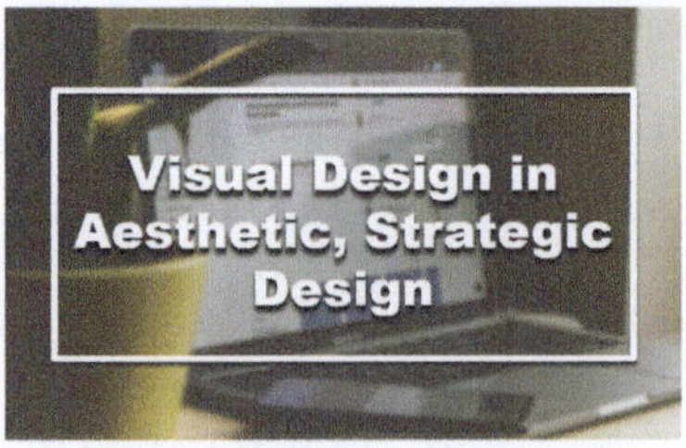

Figure 4.1: Visual design interface.

The use of visual design is to create and organize elements to

- lead the user to the functionality of the item; and
- make the style consistent.

For example, designers create and arrange the content of the website around each page's purpose and are careful enough to make sure that content provides the right visual sign. The smallest and precise details will affect the thinking and feelings of users. So, one must show the right things in the right way.

Uncertainty of users is your enemy. If users of you have to stop and think about your design if you won't gain their trust. Therefore, a visual design should have their recognition of the key aspects and maintain a balance between a design that is new and strong and something they expect to see.

The application of visual design will depend on the item, its organization, and its users (including their culture). A major aspect of visual design is that you work to accommodate user constraints like a cognitive load. For example, chunking is used to help the users easily understand and remember the information.

Successful Design = Careful Visual Design + Good Usability

As a visual designer, you should attempt to make content that is predictably organized but at the same time impressively distinct. You do this so that the users get the context they need. A visual designer wants to meet the expectations of the users to win and keep their trust, and influence them with a great brand presence. Brand style guides generally define standards about typography, color use, etc.

However, there are some principles in virtual design that are universal. For instance, whitespace and effective chunking will always be welcomed by the users irrespective of what you want to show to them. Similarly, visual design should be a factor in the application of copywriting.

So, those thoughtfully chosen simple words should appear in well-placed parts which can be easily digested by the users. You shouldn't forget the accessibility as part of the design process when you judge how to approach the best way to visual design in the context of color, contrast, etc.

4.3. VISUAL ELEMENTS IN UI DESIGN

The fundamental building blocks of any type of visual communication such as line, shape, color, texture, and pattern are formed by Visual elements (elements of art, even). They have a limited meaning alone, same as the words in the language, but we can tell the number of stories just as we start looking through them. Learning a good visual system is similar to learning of new language. To understand them properly, we need to look into them, piece by piece, till we find their most basic elements.

4.3.1. Line: The Visual Guidance

Different from mathematics, lines in design can have both lengths as well as breadth. Visual elements comprise the most flexible of them all, as it suggests many other shapes, patterns, and textures. We can use lines for splitting things from each other or to guide our eyes towards something. It can suggest conceptual ideas like time, long or short, and rhythm, straight or wavy.

4.3.2. Shapes: The Illusion of Space

One or more line forms a shape but it stands out of its environment because of its described boundaries. These shapes can step out of the 2-dimensional world to form the illusion of space and depth. The minimalism's present era makes sticking to 2D more common.

Although, whenever you put a drop shadow on a CTA button, this illusion is created by you. Geometric and organic shapes offer a different way of grouping. On the one hand, organic forms appear more natural and human to our eyes, whereas on the other the Geometry means to order and predictability, great in UX.

4.3.3. Colors: The Main Triggers of Emotion

As one would expect, this element cannot exist on its own, either. To exist, colors need a platform. Just remember that different colors don't need to trigger positive or negative but different kinds of emotions.

Think of neutral colors like black, white, gray, warm colors like shades of red which stimulate more and cool colors like shades of blue, which quickly calm us down. Colors don't just mean things on their own they also interact with each other. You can enhance your website's messages by choosing different color combinations. The four basic color schemes are: monochromatic, analogous, complementary, and triadic.

4.3.4. Monochromatic Colors

This type of scheme uses one main color with all of its shades from white to black. Many times, people use it on minimalistic websites that highlight the content itself. We can see these types of websites in E-commerce fashion brands.

4.3.5. Analogous Colors

This type of scheme uses colors that lie next to each other on the color wheel. For example, during the creation of the image of the forest, Forest uses different shades of green and blues to produce a pleasant look. To find this scheme in nature, just think of an actual forest, which explains its visual appeal.

4.3.6. Complementary Colors

This type of scheme of colors lies opposite to each other on the color wheel. These pairs make each other stronger- one looks more vivid when that one is next to its pair. Like the yellow CTA button looks extremely brilliant when compared to the dark purple background and the bold white paragraphs on this, help to make this contrast even more striking. Love it or hate it, but these types of color combinations yell for recognition.

4.3.7. Triadic Colors

This type of scheme uses three main colors which perfectly balance with each other. They create an equilateral triangle on the wheel of color. Without any surprises, it gives a calm and harmonious look to the websites with a large audience. This scheme is used by slacks with its purple-yellow-green palette. Here everything is balanced, and the elements don't overpower each other.

4.4. FIVE FUNDAMENTAL ELEMENTS OF INTERFACE DESIGN

Before embarking on any interface design project, it is relevant to discuss the concept of content like what is this? of context like who's it for? And about the audience, like where does it live? Identifying the goals, constraints, platforms, audiences, etc., will help to shape the designs and help in developing the best possible visual solution for them.

Figure 4.2: Fundamentals elements of interface design.

Before designing these solutions, there are some key fundamental elements that we should keep track of the following.

4.4.1. Language

There are different ways through which we can work with words within our project:

(i) Word as Naming/Branding

Naming is the first thing that comes to mind when we consider language as a design tool. The name of the product has the power to set the mood at the very entry-level of your brand's experience.

(ii) Word as Tone/Attitude:

It is quite useful to come up with a set of words that can be applied to each design and direction you have. Tones vary with the intentions and decisions of your brand on whether they should be strict, scientific, friendly, entertaining, fun, or playful should align with those intentions.

4.4.2. Color

Color also sets the mood for your interface as humans generally respond to them in a very emotive and subjective way.

As color is both relative and subjective, we have to find out how to control that subjectivity a little bit what we want it to do for our interface, to get it to do, i.e., create the intended atmosphere, mood or function.

(i) Color as Mood:

A person often admits the connotations with such as very serious, when we have a purely black and white creative. That stems from cultural connotations of black and white printing in newspapers and early photography. We can immediately change such connotation by adding a simple, vibrant color to the creative.

It is safe to know that some color tones come from the real world while some were manufactured. An example of a cultural convention is, Pink represents girls and blue represents boys.

(ii) Color as Navigation:

In reference to a function, color can be used in a variety of ways, but the most prominent way is the navigation in our interface.

4.4.3. Imagery

The imagery comes in different styles like shapes, illustrations, photography, 3D renderings, etc. But whatever the style is, in the context of an interface design they can serve as content, mood, and navigation.

(i) Imagery as content

Imagery can serve in two primary ways as content in our interfaces:

- *Immersive imagery:* This is generally common for games, apps, animations, etc., where the imagery exists as a design element in itself, breaking out of the box and being in its world.
- *Contained imagery:* This type of imagery is usually for delivering some sort of content to the intended audience where the imagery exists in some form of grid or frames. Examples include a grid showcasing a hierarchical category of some sort, product images in an e-commerce site search result, etc. Contained imagery comes up with a very practical comparative structure which is very functional when you need to work across content that needs to work across or that needs to be searched a lot at different sized platforms.

(ii) Imagery as mood

Imagery can be used to set a tone or ambiance with which the audience can relate to. Generally, brands do have a style guide that definite states the type of images to be used and how they should be used over an entire brand experience.

(iii) Imagery as navigation

Imagery can also serve as navigation, apart from the aesthetic function. In the e-commerce space, this is very common. For instance, on the page of Balenciaga product, imagery serves as the main navigation element. Imagery as navigation also has massive adequacy in geospatial sectors.

4.4.4. Typography

In the context of an app or a website, typography can work in many different ways. This includes working as content which we use, as part of a branding system to give your product a recognizable identity or as a part of an interface in terms of buttons or labels.

(i) Typography as content

Content consumption can be in the form of text, videos, audio, etc. But the text is somehow better than these other forms in some areas. Think of how to represent the word "ambiguous" with imagery or video. When it becomes difficult to depict certain representations typography comes to the rescue. In terms of screen real estate, typography is fast, accurate, and economical.

4.4.5. Icons

Normally, icons compete for visual interests and recognizability. One of the things on which people mostly get confused about is the difference between an icon and a symbol.

(ii) Icon vs Symbol

An icon can be seen as a visual representation of a thing that is relatively realistic in comparison to what the thing is. But a symbol doesn't necessarily look like the thing that it symbolizes. An icon is a pictorial representation, whereas a symbol is non-pictorial. It will be tough to show something as broad and something as abstract as music with a fixed single visual image.

(iii) Icon as interface

Icons and symbols act as an interface element by showing some actions. For example, the notification icon looks like a bell on medium, so it shows that and as a symbol, it stands for the concept or an idea of notifications. The magnifying glass is a very common example that is used for two different things. Its iconic function is- zoom in and zoom out and is used when we want to zoom in and zoom out, but it's also used when we type text into a search for something, which is its symbolic function.

4.5. ESSENTIAL USER INTERFACE DESIGN TIPS

The Interface design that focuses on the layout of the functionality of interfaces is a subset of user experience design, which focuses on the bigger picture: means, not only the interface but the whole experience.

4.5.1. Know Your Users

Beyond everything else, you should know that who your users are: inside and are outside. It means you should know all the demographic data which your analytics app(s) can pull. But more than that, it means you should know what they need, and what stands in the way of them to achieve their goals.

It requires a more careful analysis of stats to get to that level of empathy. It also requires getting to know the people who use your website. It means talking to them face to face, watching them use your product and maybe others, and asking them questions which go deeper, like-What do you think of this design? What are their goals? What stands in their way to achieve those goals? How a website can help them to conquer or work around those challenges? Always try to know what your users want. Explore further and find out what the users need. Beyond everything, needs grow desires. If the user's deep-seated need could be addressed by you, their wants will be addressed by you while also satisfying more fundamental requirements.

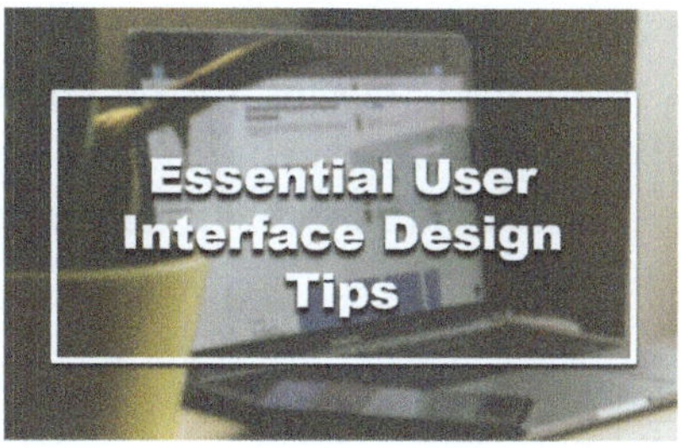

Figure 4.3: Essential user interface design tips

The insights one will discover from speaking with users and analysis of data will inform every decision you make, from how your interface is used by people to what types of content you'll highlight within that interface.

4.5.2. Define How People Use Your Interface

You need to define how people will use the interface before you design it. With the increasing prevalence of touch-based devices, it is a more central concern than you might think. Just look at Tinder: the experience of the users of this app is defined by the ease and impulsivity of a simple swipe.

People use websites and apps in two ways: directly and indirectly. Indirectly, interacting with the interface elements of the product and indirectly by interacting with user-interface elements which are external to the product.

(i) Examples of direct interactions:

- Button tapping;
- Card swiping;
- Drag and drop of an item with a fingertip.

(ii) Examples of indirect interactions

- Point and click from the mouse;
- Using commands/shortcuts key;
- Typing in the form field;
- Drawing on a Wacom tablet.

4.5.3. Set Expectations

There may have some consequences also during the interactions with a site or an app: erasing a website, the clicking of the button can mean spending money or making a disparaging comment about grandma's birthday cake, etc. And any time consequences are there, anxiety also there.

So, you must assures that the users know what will happen if they click that button before they do it. This can be done by you through design and/or copy.

(i) Setting expectations with design

- Highlighting the button that corresponds to the desired action.
- In combination with copy use a widely understood symbol such as for a delete button use trash can, to add something use plus sign, and for searching use magnifying glass.

- Picking a color with a relevant meaning like green for a "go" button and red for "stop."

(ii) Setting expectations with copy

- Write the button copy.
- In empty states, provide directional/encouraging copy.
- Asking for confirmation after delivering warnings.

It makes sense to ask people if they are sure about the actions with irreversible consequences, like deleting something permanently.

4.5.4. Anticipate Mistakes

To err is human; to forgive, divine.—Alexander Pope, "An Essay on Criticism"

People make mistakes, but they shouldn't always have to suffer the reactions. To lessen the impact of human error, there are two different ways:

- Try to prevent errors before they got happened.
- Provide ways to fix them after they got happened.

In eCommerce and form designing, you can see a lot of mistake-prevention techniques. Until you fill out all fields, the buttons remain inactive. Forms detect that an email address hasn't been entered properly. It can ask questions through Pop-ups like if someone wants to abandon the shopping cart.

The expectation of errors is less frustrating than trying to fix them after the fact. This is because they generally occur before the satisfying sense of completion which only comes after clicking the "Next" or "Submit" button.

It is said that sometimes you just have to let accidents happen. That's when detailed error messages come into.

Make sure they do two things when you write an error message:

- Explain the problem. Like, "You said you were born on Mars, but humans haven't colonized yet."
- Explain how to fix this problem. Like, "In the column of the birthplace, please replace mars with Earth."

The Poka-yoke principle is the principle of anticipating user error. It is a Japanese term that means "mistake-proofing" or "inadvertent error prevention."

4.5.5. Give Feedback—Fast

In the real world, a person gets a response from the environment. An individual speaks and others respond naturally to it. Like a person scratches a cat and it purrs or hisses depending on its mood and how much we suck at cat scratching.

Digital interfaces fail to give much back and leaving us wondering whether an individual should reload the page, restart the laptop or just throw it out from the nearest available window.

4.5.6. Stick to the Brand Guidelines and the Tone of Voice

How you can use visuals is determined by different industries. We base our judgment on the credibility of a site after scanning the home page for a few seconds. Would you trust a law firm on whose website, crazy textures and pink illustrations are used? No, you would have concerns about it. Naturally, you can enhance it, but you have to position your brand in the same category as your competitors have. The company's industry and its values should always be respected by visuals.

4.5.7. Always Think in Systems

Visual consistency has the key as it helps your users to navigate from your site and remember your brand. It also indicates its quality. You probably offer a well-designed product, if you have taken the time to consider every small detail on your website.

Consistency also means that you should choose them selectively when confronted with endless combinations of visual elements. Stick with outlined icon set, if you use it. Use shades of two or three main colors. A new landing page doesn't necessarily give a good reason to introduce the third one. These things may look unimportant, but they add up.

4.5.8. Make Sure That Content Comes First

To transfer the brand message better, visuals help a lot. Even though visuals come as a cool benefit, one must make sure that visitors will stay on your site to read your content so, never use them just too pretty up your site. When you can tell something (complex idea) in writing, an illustration can show it.

Visuals can express a feeling of your brand's personality, give a deeper understanding of your product, and help to communicate your tone of voice. But users still need to find what they are looking for. Give your content to people so that they value it. Written content and visuals both play an important role, one cannot perform well without the other.

4.5.9. Respect Negative Space

We call negative or white space to the area between the elements of a design. Keep one thing in mind: White space does not mean wasted space. So, never let the word "white" trick you; make it any color that your heart desires. People are always in a hurry and want to find things as soon as possible.

If a person crams every bit of info about the product into one page, nothing will stand out. So, always try to guide the visitor's eye to the most important information you want them to understand and eliminate other unimportant factors. Negative space allows tired eyes to rest a bit while the brain processes the new information.

4.5.10. Experiment

Visualizing something can have multiple good solutions, and visual elements can prove very versatile. Never feel shy in trying out more than one solution, just make sure to test which one performs better. It could wind up a small thing like a landing page and also how you position your brand. Flexibility gives us a very recent example.

Even some new logos shocked the design community all around the world. It gives a new design to a website as well. But plenty of things stayed the same, like colors, the friendly tone of voice, the simple navigation system, and focus on people. Slack is experimenting with growing up.

4.6. PRINCIPLES OF VISUAL DESIGN IN USER EXPERIENCE

Generally, one can immediately say whether it is appealing or amiss after looking at the visual. As in Don Norman's model of emotional design, they often play out at the deepest level. Nevertheless, few can verbalize why a layout is visually attractive. Graphics can drive engagement and increase usability, which takes advantage of the principles of good visual design.

Visual-design principles tell us how design elements like line, shape, color, grid, or space go together to generate well-balanced and thoughtful considerate visuals.

This topic defines five visual-design principles that impact UX:

- Scale;
- Visual hierarchy;
- Balance;
- Contrast;
- Gestalt.

4.6.1. Scale

The principle of scale is generally used, and almost every good visual design takes advantage of it.

Definition: The principle of scale refers to using relative size to indicate its significance and rank in a construction.

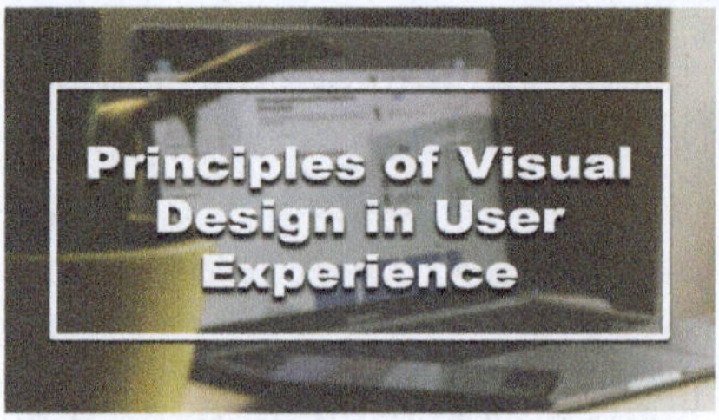

Figure 4.4: Essential user interface design tips.

It means the most important elements in a design are bigger than the ones that are less important when this principle is used properly. The reason is simple behind this: it's more likely to be noticed when something is big.

A visually pleasing design generally uses three different sizes, not more than that. Having a range of different sized elements will not only create variety within your layout, but will also establish a visual hierarchy. Be sure to emphasize, mark the most important aspect of your design biggest and follow the hierarchy according to their usage.

When the principle of scale is used properly and the right elements are emphasized, users know how to use it and will easily parse it.

4.6.2. Visual Hierarchy

A layout with a good visual hierarchy will be easily understood by your users.

Definition: The principle of visual hierarchy refers to guide the eye on the page so that it attends to different design elements in the order of their importance. This means it is used to rank design elements and influence in the order you want your uses to view them.

Visual hierarchy can be implemented through variations in scale, value, color, spacing, placement, and a variety of other signals. You can establish each element in its rightful place and help the most important elements stand out.

It controls the delivery of the experience. If its layout is missing a clear visual hierarchy, you have a hard time figuring out where to look on a page.

To create a clear visual hierarchy, use 2–3 types of face sizes to indicate to users what pieces of content are most important or at the highest level in the page's mini-information architecture. Or, consider using muted colors for less important items and bright colors for important ones.

Visual hierarchy can also be defined by scale, therefore include various scales for your different design elements. A general rule of thumb is to include small, medium, and large components in the design.

This is the arrangement of elements (mainly text) in a fashion to catch the reader's eye on the main elements first and then the less important one and then the lesser important than before.

4.6.3. Balance

Balance is just like a seesaw: you are balancing design elements rather than weight.

Definition: The principle of balance refers to the proportion of design elements or a satisfying arrangement. Balance takes place at the time when there is an equal distribution (but not necessarily symmetrical) of the amount of visual signal on both sides of an imaginary axis passing by the center of the screen. Generally, his axis is vertical; however, it can also be horizontal.

Just like when balancing weight, the design would feel a bit unbalanced in case you were to have one small design element and one large design element on the two sides of the axis. The area taken by the design element is taken into consideration during the time of creating balance, not just the number of elements.

The imaginary axis that you establish on your visual will be considered as the reference point for the way you can organize your layout and will help you to know about the present state of balance on your visual. No one area draws your eye so much that you can't see the other areas in a balanced design (even though some elements might carry more visual weight and be focal points). Balance can be the following:

- Symmetrical: In this, elements are distributed symmetrically relative to the central imaginary axis.
- Asymmetrical: elements are distributed asymmetrically relative to the central axis.
- Radial: elements radiate out from a common and central point in a circular direction.

One can choose any type of balance in your visual based on the matter you want to convey. Asymmetry is engaging as well as dynamic. It creates a sense of movement and energy. Symmetry is static and quiet. The radial balance will always take you towards the center of the composition.

4.6.4. Contrast

This is one of the commonly used principles that make specific parts of your design stand out to your users.

Definition: The principle of contrast denotes the juxtaposition of visually dissimilar elements to convey the fact that these elements are diverse (e.g., have different functions, belong in different categories, and behave differently).

In simple words, to emphasize that they are distinct, contrast provides the eye with a noticeable difference (e.g., in size or color) between two objects (or between two sets of objects).

Often, the principle of contrast is applied with the help of color. For instance, the red color is frequently used in UI designs, especially on iOS to signify deleting. The bright color signals that a red element is different from all other elements.

In UX, the word "contrast" often brings to mind the contrast between text as well as its background. Sometimes, to deemphasize less important text, designers deliberately decrease the text contrast. However, this approach is dangerous since reducing text contrast also reduces legibility and it may make your content inaccessible. To ensure that your content can still be read by all your target users, color contrast should be used.

It refers to differentiating one item from another. It can be done in different ways other than just using colors in contrast. It can be done by using completely dissimilar text, dissimilar sizes as well as by using contrasting colors. The objective of this is to catch the reader's eye or create a focus on something important.

4.6.5. Gestalt Principles

Gestalt psychologists introduced this set of principles in the early twentieth century. They capture the way human beings make sense of images.

Definition: Gestalt principles describe the way humans organize as well as simplify complex images that consist of many elements, by subconsciously arranging the parts into an organized system that creates a whole instead of interpreting them as a series of disparate elements. In simple words, principles introduced by Gestalt capture our tendency to perceive the whole as opposed to the individual elements.

There are some Gestalt principles, including continuation, similarity, closure, common region, figure/ground, proximity as well as symmetry and order. Proximity is very necessary for UX. It denotes the fact that items that are visually closer are perceived as part of the same group.

4.6.6. Proximity, Similarity, and Continuity

It involves showing visual spaces to express the relationship between as well as among the content. Similar things are just grouped. When elements of the design are not grouped, they express the lack of relationship. In case she is in a hurry or quick browsing mode, this makes the eyes of users less strained to understand and filter information.

4.6.7. Reification and Closure

Objects can be identified by people even if most of the parts of them are absent or unaccounted for. This is known as reification. In the next topic, the importance of closure is evident as it lets people see a white square rather than broken circles that are four in number. The brain fills the gaps that are present, and also it matches what a person perceives with amicable patterns that are hoarded in the memory of an individual.

4.6.8. Figure-Ground Relationship and Relative Size

The heading clearly states that a structure or ground is always present and there is always a figure over it. These figures and grounds can be swapped to show items that a person wants to show or convey, this proves to be a cunning move.

The figure and grounds are held and discerned following the viewer, they can transpose as per the viewer. The second and third images that are illustrated below can be viewed in two ways, for instance, several people will see the pair of faces, whereas a few of them will view it as the thin vase.

The differences between the perceptions of these people are the relative size of the vase and the range or the gap that lies in between the faces. This increasing distance begins to intensify to a much more extent on the vase instead of the faces.

4.6.9. White Space

White spaces do not signify the white-color spaces. They refer to the space that is evident on the canvas where there are no traces of the element are found. The white spaces are often named as negative spaces, that assist designers to focus on the essential information and set apart dissimilar information at the same time. White space is the paramount and key design principle or proposition.

There is a video available at the end of the chapter which clearly shows how the white space proves to be significant and a key factor. The picture that is given below (contrast section), portrays how white spaces dispose of the elements and gives space to the layout or the structure.

4.6.10. Alignment

Alignment is a significant factor that is identified by humans in all phases of their life. Starting from the architectural structures, clothing, arrangements of room, food display, various designs of hardware, or a layout design that has been discussed and mentioned here.

The use of alignment is that it provides a look that is tranquil along with improving the clarity of the page. This also gives assistance and guidance for grouping the text logically to foreground the hierarchy of the elements.

By being consistent, one can follow this rule easily which is an essential method.

4.6.11. Why Visual-Design Principles Are Important

Often the question comes up that why should people care about and understand visual-design principles? Apart from making something "look pretty," understanding or a proper apprehension and taking advantage of them serves to:

(i) *Increase usability.* By following these visual-design principles sometimes results in layouts that are effortless and unchilling to use. For instance, the golden ratio, which is used time and again for generating beautiful works of art, was also made us of in typesetting to create a graphically agreeable relationship between font size, line height, and line width.

The outcome of it, typically led to shortened line lengths, which established balance (via white space) on a webpage, and through this, it was easier to read texts. There will be a rapid increase in success rates and user engagement of visual design when it is put together with a strong interaction design.

Figure 4.5: The importance of visual design principles

(ii) *Produce emotion and delight.* Aesthetic things bring out positive emotions. (As a matter of fact, the aesthetic–usability effect states that the visually tempting design found by the users, may be more lenient of small usability troubles). Designers can create aesthetic and appealing UIs by following the proposition of good visual designs which eventually users enjoy and feel optimistic about.

(iii) *Strengthen brand approach.* The user trust, interest in the value of products is built by creating a powerful visual system. The strong visuals not only create trust and interest but also accurately represent, promotes, and strengthen the brand.

4.7. APPLYING COLOR THEORY TO DIGITAL DISPLAYS

The background of the text should be in bold colors, the use of textures and patterns must be made less as it makes the letterforms difficult to differentiate, and it also makes it not clear enough to read; hence the meaning of the text is left undeciphered. The combinations and background of the text colors must be chosen properly and keenly. The value contrast that is between the body of the text and the background colors should be set at the least 80%.

4.7.1. Contrast with a White Background

The maximum value contrast comes from black text on the white background and excellent readability for body text where the essence of the text is not lost. The contrast between charcoal gray (#333333) text and white background is about eighty percent, which provides good value contrast at the minimum.

The following dark colors provide good to excellent contrast and legibility for text on a white background:

Excellent: Black (#000000) text on a white background.	Excellent: Kashmir green (#003300) text on a white background.	Excellent: Midnight blue (#000033) text on a white background.	Excellent: Burnt umber (#330000) text on a white background.
Good: Charcoal gray (#333333) text on a white background.	Excellent: Peruvian turquoise (#003333) text on a white background.	Excellent: Deep purple (#330033) text on a white background.	Excellent: Raw umber (#333300) text on a white background.
Good: Slate (#333366) text on a white background.	Very good: Forest green (#006600) text on a white background.	Good: Navy blue (#000066) text on a white background.	Good: Deep burnt sienna (#660000) text on a white background.
Good: Indigo blue (#330066) text on a white background.	Very good: Viridian green (#006633) text on a white background.	Good: Prussian blue (#003366) text on a white background.	Good: Deep burgundy (#660033) text on a white background.

4.7.2. Contrast with a Black or Dark Background

On a black background, the high-chroma colors yellow (#FFFF00), green (#00FF00), cyan (#00FFFF), and magenta (#FF00FF) proves to be an excellent match. On the other hand, white text used on a black background gives a value contrast that is very high and stinging, it is less readable and incomprehensible also causes eye fatigue to a great extent as compared to the black text on a white background. All light-colored text on dark backgrounds irritates the eye and eye fatigue. Generally, the use of light colors on dark backgrounds for small amounts of bold text must be put aside—for example, headings, labels and links in navigation bars and menus, button labels, or pull-quotes—and make sure that the font size is sufficient to validate readability and the quality or the clarity is preserved.

The table mentioned below is the high-chroma colors that provide very good contrast and legibility for text on a black background:

Very good: Yellow (#FFFF00) text on a black background.	Very good: Green (#00FF00) text on a black background.	Very good: Cyan (#00FFFF) text on a black background.	Very good: Magenta (#FF00FF) text on a black background.

4.7.3. The Problem with Blue

Because only two percent of all retinal cones—the photoreceptors that ensure color vision—are blue-sensitive cones and the eye draws attention towards blue and brings it into focus in front rather than on the retina, visual acuity for the blue range of the spectrum is poor and it keeps on reducing with age.

The most common problem with the color blue is that when the symbol identification is done, the performance of the user is very bad with the symbols that are blue as compared to the symbols of various other colors.

Consequently, the use of dark colors or highly saturated shades of blue for text or other elements consisting of fine lines on a black or other dark background must be avoided as much as possible. Along with this, the text or symbols consisting of fine lines in highly saturated blue hues on a white or another light background must also be avoided. Text in desaturated blue hues is appropriately legible and readable on light backgrounds. Blue is considered a good background color.

4.7.4. Contrast and Legibility

To provide the best legibility, it should be made sure that text contrasts suitably with its background in both hue and value. When there is an inadequate contrast between the hue or value of text and its background color, the text loses its clarity and it appears blurred or has a halo effect around it which shows that the readability of the text is lost, making it strenuous to read and also causes eye strain and irritation.

Text that is in a color that contrasts well with an achromatic background of black, gray, or white or black or white text on a high-contrast, color background most of the time gives better legibility in comparison to when both text and background are in different chromatic colors—unless the two colors contrast greatly in both hue and value.

The following table depicts the color combinations which provide good contrast and legibility:

Good: Black (#000000) text on a cyan (#00FFFF) background.	Very good: Black (#000000) text on a yellow (#FFFF00) background.	Good: Black (#000000) text on a pale yellow green (#99FF99) background.	Good: White (#FFFFFF) text on a blue (#0000FF) background.

The use of gray color must be done very less or it would be better when avoided for text or symbols which consists of fine lines on a color background, reason for this is a successive-contrast effect causes them to take on the hue of their background color's complement.

All qualities of Web-safe grays are excessively dark to be made use of as effective background colors for the body of text that is black. However, the non-Web-safe light gray (#EFEFEF) creates a good background color for black text. The white text contrasts adequately with a charcoal gray (#333333) background.

4.7.5. Enhancing the User Experience with Color

The following design principles are implemented concerning the use of color to amplify the experience of a user regarding an application program or Web site:

(i) Use color consistently throughout

Before designing an application or Web site, creating the design guidelines are mandatory, including conventions and understanding for the use of color. These conventions should convey all purposes for which an individual should use color, what colors apply to different and particular types of elements, and the meanings and reasons connected to specific colors.

These color conventions should be used consistently and regularly throughout an application or Web site. When users understand and interpret properly the meanings behind the colors, then they will start applying and using those meanings in whatever place they come across the colors.

If an individual's use of color is reliable and uniform, then users will face difficulty in constructing a mental model of color usage, or they will be unable to comprehend the meanings of specific colors accurately.

(ii) Use color both to support users' tasks and for branding

Ideally, to make sure maximal user performance in applications, it is an elite way to use color only to back users' activities and decision-making. Because color is such a noticeable, influential visual cue, its arbitrary use for the aesthetic role can reduce the performance of the user. Though, on the Web, the use of color for purposes of branding is highly imperative.

On the Web, color branding offers a sense of place, which is highly critical to user orientation and assists in preventing users from becoming lost in hyperspace. Try to confine the use of color that is exclusively for branding to certain areas of Web pages – for instance, a header, an emblem in the upper-left corner of a page, or a footer.

In the core content area of a Web page, in an interactive Web application, or a Web form, or, the use of color should be related to the task. In such settings, use color cautiously – only where it offers real value and enhances the performance of the user. Excessive use of color lessens its efficiency.

(iii) Use color to increase user satisfaction

Use color to improve the aesthetic appeal of an application or Web site – even if the use of color does not further the performance of the user. It is generally seen that majority of people prefer color instead of achromatic Web pages, and on the Web, user satisfaction is just as imperative as user performance. Though, ensure that the use of color is not visually disturbing, worsening usability and reducing users' potential to finish their tasks effectively.

4.7.6. Using Color for Identification, Grouping, and Emphasis

It is generally seen that there are several ways in which you can use color to assist in identifying objects, guide the attentiveness of users, communicate structure, or indicate status. Use color to do the following:

(i) Identify groups of related Web pages

One can make use of distinctive color schemes to recognize groups of related Web pages, communicating a Web site's high-level information architecture and making it simpler for users to rapidly recognize the part of a Web site to which a page belongs.

(ii) Distinguish specific Web pages

One can make use of a distinctive color scheme to differentiate specific Web pages. Doing so offers a sense of place and, thus, can help in reducing user errors and confusion. For instance, a purchasing process funnel might have a distinctive color scheme from the rest of an e-Commerce site.

(iii) Group controls or information in a window or on a Web page

One can make use of background color to set off groups of allied controls or related items of information that are adjoining to one another. One of the main benefits of using color for this purpose – instead of white space, borders, or some other visual cue – is that color does not expend any extra screen real estate.

4.8. DESIGNING FOR SENIOR CITIZENS ORGANIZING THE WORK SCHEDULE

"At the time of designing for older adults – that's what we call them because not everybody prefers being thought of as a senior citizen – contrast is key," answers Dana. "So, when while considering using colors next to each other, it is important to remember that many people can't see distinctive hues or don't see the differences in similar colors – like green and blue – because of a loss of visual acuity that comes with age. If you want to put text on a colored background, make sure there's plenty of contrast."

"As for fonts, sans serif fonts are best," endorses Dana. "Older adults and people with low vision have less trouble processing typefaces like Helvetica or Arial. Without the series, it's simpler to identify characters. The thing one will hear the most from older adults, though, is to make the type larger.

Young Web designers usually start their designs with justly lesser types. The thinking is that anybody can enhance the size through the browser, the majority of the adults of older age don't know how to do that. So, go with 11-point or 12-point type, with plenty of leading, or line-height."

Caroline agrees: "For colors, there is a need for good contrast between the foreground color and the background color. Use large, clear fonts. For both fonts as well as color, it is imperative to make sure that your choices are adjustable, and make it very tranquil for users to make those adjustments."

4.8.1. Good Legibility Benefits Everyone

"There is the latest trend in the design of Web sites and applications to design content using very small fonts and, usually, very low contrast as well," remarks Pabini. "I've devised the term the Wireframe School of Design to define this trend, as the design of these sites believes to be as colorless and spare as a wireframe.

While straightforwardness in visual design is a commendable goal, these young designers are offering things too far. They're both young and have decent eyesight, because, else, they'd realize their text is unreadable to many people – and not just to older adults, but also to the majority of the people of all ages whose eyesight is weakened or who have poor visual acuity.

- Of all Americans on the Web, 11% are now of more than age 64.
- Boomers (aged 50–64) directed the US population, making up 32.5% of the adult population and 36% of the adult online population and accounting for about 33% of daily Web traffic.
- Over 4 million Americans have low vision – that is, the impaired vision that neither medical treatment, corrective lenses, nor surgery can fully restore. Of these people, 68% are over the age of 65.
- More than 50% of the US population need corrective lenses to rectify some vision problem.
- Nearsightedness or myopia affects more than 25% of the US population.
- More than 25% of all American school-age children have a vision-related problem.
- Approximately 4% of people of European origin either are color-blind or have a color-deficient vision – particularly, about 8% of men, but only 0.4% of women.

"Extrapolate these numbers to the global population, and one will rapidly realize that poor legibility of text on Web sites influences a lot of people negatively."

4.8.2. Understanding the Needs of Older Adults through Research

"In 2004 and 2005," says Dana, "I indulge in some work with Ginny Redish and Amy Lee for AARP, in which we advanced a set of heuristics for intending Web sites for older adults: 'New Heuristics for Understanding Older Adults as Web Users.'

Then we used the heuristics to review 50 Web sites from the perspective of older users, recording our findings in 'Designing Web Sites for Older Adults: Expert Review of Usability for Older Adults at 50 Web Sites.'"

Caroline endorses following Heuristics 14 and 16 from "AARP audience-centered Heuristics: Older Adults," saying, "This document is dependent on broad research and a literature review Dana Chisnell and Ginny Redish conducted." Dana also proposes some publications from the National Institute on Aging (NIA) and the W3C.

4.9. CREATING A UX DESIGN STYLE GUIDE

As Nathan Curtis labeled on Twitter, a style guide is a document that a UX designer crafts to document a rising and ever-evolving set of design guidelines that ascend from the design process. In creating a style guide, UX designers are chiefly recording their thought process as they design a Web site, application, or system.

Figure 4.6: Creating a UX design style guide.

In this way, the crux of making a style guide is verifying your own design decisions. Who is the audience for this document? In this section, the main area of focus will be on several crucial questions about style guides to help UX designers create effective documentation.

4.9.1. Documenting Your Color Palette and Typography

A Web site or application's color typography and palette are archetypally the first things that get frozen, initially in the visual-design process. These are imperative design decisions that nearly everyone on a product team will have a view about.

4.9.2. Specifying Your Color Palette

Depending on how many parts one needs to embrace in their style guide, the documentation of the color palette might comprise of the resulting optional and mandatory sections:

(i) Mandatory sections:

- "Text Colors"
- "Primary Color Palette"

(ii) Optional sections:

- "Tertiary Color Palette"
- "Secondary Color Palette"
- "Palette of Grays"

4.9.3. Defining the Background Color

Describe the color of the background, even if it is white. Gray or white background with shadows on elements usually produces a layered look for an application. But it is highly important to be careful in selecting a shade of gray for your background. Making too much gray in your application can make the design seem dull or out of date. In addition, a background in the wrong shade of gray can form value-contrast issues, creating text hard to read.

(i) Choosing Text Colors

It is important to take care in defining each color in your user interface, containing text colors. While working on initial designs, it's tranquil to get carried away with colors, particularly those for text. But there is a need to use separate colors for the following types of text:

- system messages such as warnings and errors;
- paragraphs of text;
- links.

Restricting the color palette for text in application to no more than three or four colors. This will not only help in making the design appears to be more consistent but also eradicate a lot of overhead for the Development team in creating a site's CSS.

4.9.4. Designing Typography

Knowing the right font to pick for a Web site or application in the field of the most skillful visual designers, who comprehend branding. But typography design is not confined only to have awareness about what font to use.

There is also a need to be vigilant about that font is readable, legible, and aesthetically attractive across distinctive platforms. This needs complete knowledge of all the devices and platforms on which your application or Web site will be intended to be used. For legibility, it is recommended to use a font size of not less than 12 to 14 points.

4.10. CONCLUSION

In the conclusion of this chapter, the basic concept and ideology behind the visual interface design have been discussed. In this chapter, the importance of visual design in aesthetics and strategic design has also been discussed. This chapter discussed the various visual elements that play a significant role in UI design such as line, shapes, colors, monochromatic colors, analogous colors, complementary colors, and triadic colors.

This chapter also discussed the various fundamental elements of an interface design which include language, color, imagery, and icons. In this chapter, the various types of user interface design tips that play an essential role in designing such as know your users, define how people use your interface, set expectations, anticipate mistakes, give feedback—fast, and respect negative space has been discussed.

This chapter also discussed the different principles of visual design in user experience which include scale, visual hierarchy, balance, contrast, gestalt principle, proximity, similarity, and continuity, reification, and closure. Towards the end of the chapter, how to create a user experience style guide has been discussed which emphasizes documenting the color palette and typography, specifying the color palette, defining the color of the background, as well as designing the typography.

REFERENCES

1. Deepak, S., (2020). *Principles of Visual Design.* [online] Medium. Available at: https://uxdesign.cc/principles-of-visual-design-fb49004f5a67 (accessed on 6 July 2021).
2. Janet, M. S., (2010). *Designing for Senior Citizens | Organizing Your Work Schedule: UXmatters.* [online] Uxmatters.com. Available at: https://www.uxmatters.com/mt/archives/2010/05/designing-for-senior-citizens-organizing-your-work-schedule.php (accessed on 6 July 2021).
3. John, M. W., (2021). *10 Essential UI (user interface) Design Tips |* Web flow Blog. [online] Web flow. Available at: https://webflow.com/blog/10-essential-ui-design-tips (accessed on 6 July 2021).
4. Kelley, G., (2020*). 5 Principles of Visual Design in UX.* [online] Nielsen Norman Group. Available at: https://www.nngroup.com/articles/principles-visual-design/ (accessed on 6 July 2021).
5. Oluwatobi, A., (2018). *5 Fundamental Elements of Interface Design.* [online] Medium. Available at: https://uxdesign.cc/5-fundamental-elements-of-interface-design-7d8b8be47044 (accessed on 6 July 2021).
6. Gabriel-Petit, P., (2007). *Applying Color Theory to Digital Displays: UXmatters.* [online] Uxmatters.com. Available at: https://www.uxmatters.com/mt/archives/2007/01/applying-color-theory-to-digital-displays.php (accessed on 6 July 2021).
7. The Interaction Design Foundation, (n.d). *What is Visual Design?* [online] Available at: https://www.interaction-design.org/literature/topics/visual-design#:~:text=Visual%20design%20aims%20to%20improve,user%20experience%20and%20drive%20conversion (accessed on 6 July 2021).
8. Vineet, R. S., (2019). *Creating a UX Design Style Guide: UXmatters.* [online] Uxmatters.com. Available at: https://www.uxmatters.com/mt/archives/2019/11/creating-a-ux-design-style-guide.php (accessed on 6 July 2021).
9. Zsofia, P., (2019). *UI Elements: 5 Tips to Make Your Website More Visually Appealing.* [online] UX Studio. Available at: https://uxstudioteam.com/ux-blog/visual-elements/ (accessed on 6 July 2021).

CHAPTER 5

PATTERNS AS TOOLS FOR USER INTERFACE DESIGN

CONTENTS

User interfaces which are known as UI, are elements that can be used again and are repeated over time. Designers put them in wide dimensions of cases, and they should modify each to the particular circumstances of use. They are recurring solutions to common problems.

They keep away from remaking the wheel and construct an authentic solution to an interface design problem rapidly. User Interface patterns are like sustenance of the user interface design. User interface design patterns are planted in architecture and programming and they were designed to enhance the user interface development process.

The patterns that are used in user interface designs are omnipresent. They are available on sites and apps and are amicable to users. Users easily understand how to deal with an interface designed using common user interface design patterns. Through embracing the techniques of common patterns, one can grasp this knowledge and then can enlarge the use of the product.

Specific patterns that are chosen to be used must be adapted to a particular use case. User interface design patterns provide a common structure on which people should work and transform it, add or subtract parts from it till a particular set of needs that are distinctive to each design project.

5.1. INTRODUCTION

The rules and directions have always helped in capturing the knowledge regarding the design; it also assists designers to utilize that knowledge while designing user interfaces. The knowledge related to the design helps the designer to make correct decisions and stops the designers from making the same errors and faults all over again. To put in the guidelines doesn't come without challenges.

These guidelines are several and diverse, so applying them to a particular design problem becomes strenuous. In addition to this, the designer faces the conflict of designs, it becomes difficult for them in solving the design problems. The guidelines are dense, and their accuracy perpetually relies upon a 'context.'

It becomes hard while designing usable systems, to achieve this designer, need effective tools that are functional. The constructive design tools must be built on the demonstrated knowledge of design. Storing the information related to the successful design of usable systems becomes crucial for beginners as well as designers who are experienced.

The knowledge that the designers hold is elaborated in the guidelines. But as stated earlier, these guidelines include issues related to selection, validity, and applicability. One of the fine solutions related to the problem of guidelines is patterns. They emphasize the circumstances of a problem and solution, by leading and guiding the designer in making the use of design knowledge.

The patterns that are used in architecture or software engineering are not similar in formation, however, user interface design also needs its structure for patterns, usability being its focal point. Patterns used in user interface design should be structured, for the sake of being effective and efficient and be useful tools for designers.

The use of patterns as tools in a user interface is that it extensively focuses on contexts, and through this, the designer becomes aware of the when, how, and why the solution must be applied. So, instead of guidelines, patterns are most beneficial and powerful, as compared to guidelines as a tool for designers.

In software construction, patterns have become very popular and useful, as they are inspired by the work of Alexander. The use of patterns as tools in user interface marks its importance back to the times of 1994, yet during recent times an established set of patterns has still not come up.

Efforts have been made for the establishment of patterns, but still, there appears to be the absence of agreement regarding the use of patterns as UID, i.e., User Interface Design should be written down that cornerstone how they should or shouldn't be structured and framed.

Therefore, an interesting pattern language for User Interface Design hasn't emerged, although it is certainly paving its way by the establishment of a large body of patterns. And patterns prove to be more beneficial than guidelines.

Figure 5.1: Patterns as tools for user interface design.

While using patterns as tools instead of guidelines, there is a huge comparison between the two. Patterns are decisively better tools than guidelines as patterns are directly related to the context and revolve around problems. They are theoretically correct as producing patterns for user interface design is not an easy task.

Patterns for user interface design are not certainly structured in the way an architecture pattern is designed; it becomes necessary to search for a format that has been designed for user interface design and has the correct information on the crucial issues for user interface design. For the patterns present in user interface design, suitability for recounting the issues associated with usability is the foremost complication, in respect of user interface design patterns.

5.2. WHAT IS A USER INTERFACE DESIGN?

User interface design uses patterns due to which structure and order fuel combined so that it can give strong experiences to the users. The structure and order go hand in hand, in addition to the truth that the habits that people who are used to these habits find it hard to leave mainly while talking about the web. Therefore, it is mandatory for the designers that they examine user interfaces keenly before setting any final design.

Designers create powerful visual designs by using various colors, fonts, sketches, etc. In large firms, these designs are increasing rapidly over the past three years. These design languages are getting famous and consistent. According to the point of view of customers, User Interface is considered very crucial in many organizations that have different visual identities.

The products that are established must be of higher value so that users don't pay attention to how they reach from one point to another. The failure of setting this may result in such user interfaces that are hard to steer. It wastes a large amount of time for users who spend way too much time decoding the display instead of investing that time into something fruitful.

Throwing some light on accessibility helps people with disabilities to have a beautiful online experience. For instance, for people with visual impairment, the patterns are designed differently so that they can have easy access. This is possible by providing larger fonts, bolder colors, high contrasts. Working in the interest of people adds value to their life.

Figure 5.2: Picture depicting user interface of an electric car.

The user interface is a set of conversations about images, fonts, tools, colors, and where visual identity happens. Digital industries provide specialization in user interface and UX, i.e., User Experience. To enhance the business more and more, a better choice of images, colors, fonts, visuals are required which have a great impact on the business.

A design language can be a set of consistent visual elements that can be reused. While exploring a website, people look up for a proper format, so it becomes necessary that things are arranged in the correct order. This helps in obtaining the knowledge of the website without losing the essence. By plumping for the right user interface design patterns is important to take the benefit of the natural pattern, i.e., spotting.

5.3. WHAT IS A PATTERN FOR USER INTERFACE DESIGN?

Patterns are related to the elements that repeat themselves, the appearance of problems, and then finding their solutions. In the words of Alexander, "Each pattern is a three-part rule, which expresses a relation between a certain context, a problem, and a solution."

He further explains the nature of the pattern," Each pattern describes a problem which occurs over and over again in our environment and then describes the core of the solution of that problem, in such a way that you can use this solution a million times over..." these statements signifies that the nature of the pattern is very practical.

Patterns portray examples of designs that are good instead of obscure principles and policies. In addition to this, as demonstrated, these patterns are not theories or hypotheses. The pattern must contain logic and reasons on the issue of why the solution works and also to prove by stating different instances where the pattern was conveniently put to use.

Figure 5.3: Patterns for user interface design.

The nature of patterns is absolute, which assists the designers to establish new illustrations. The essence and core of a solution must be represented through patterns. For distinct purposes, the templates are not usually the same also for specific purposes, there is a requirement of an adaptation.

There is a network of patterns when referring to the collection of patterns given by Alexander and in the SE collection of patterns. Chains of patterns are used to link patterns with each other. According to Mahem, there are disparate kinds of patterns for the user interface design.

Their intensities of dealing with the problems are different. For instance, certain patterns handle small problems that deal with a single screen, on the other hand, others emphasize high or extraordinary principles like selecting the direct manipulation.

This ranking is done, or in other words, this hierarchy is set so that when sufficient patterns are recognized, the variations can be easily made. The collection which is constructed must be grounded on how the use of patterns that are in practice.

Talking about the collection of patterns for user interface design, it is evident that it has marked its existence for some years, yet they are not accessible extensively. In recent times, there is an availability of two patterns, the first one was assembled by Jennife Tidwell, which consists of sixty patterns. The second one was assembled by the Usability Group of the University of Brighton and consists of a dozen patterns. There exist huge variations in both the collection of patterns, precisely the formats used in these collections vary.

The collection of patterns given by Tidwell is constructed based on standard aspects like name, problem, context, examples, and forces, while the Brighton collection is not well structured. The similarities between both collections are that they contain anecdotes that describe the pattern. Few patterns are descripting of guidelines. The rest of the problem lies in the core of patterns.

The problem description and the solution. The description of these problems is very obscure, as a result, attention is shifted towards the solution. But the purpose of patterns contradicts this, the verification of the problem must be done correctly so that the solution solves the real problem.

5.3.1. Difference between User Interface and User Experience

UX refers to the User Experience which is supposed to meet the requirements of customers, without any hurdles. The other thing is the way the products are produced, by simplicity and elegance which provides delight to the customers while using them.

The experiences that users get are true in the way when customers get what they demand or ask for. To cater to the needs of the customers by giving them high quality of user experience, the services are integrated into complex disciplines, which include engineering, marketing, graphical, industrial, and interface design.

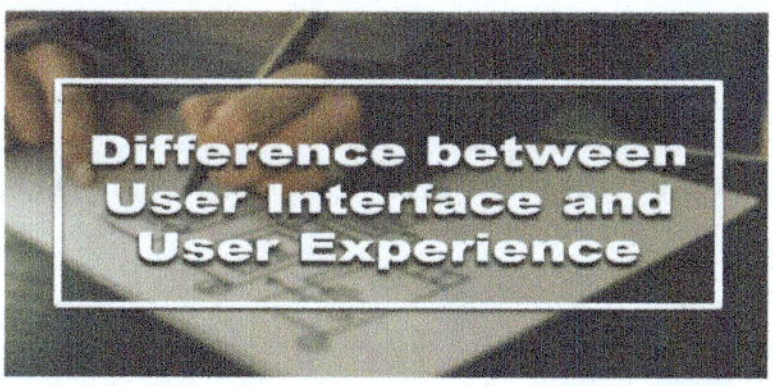

Figure 5.4: Difference between a user interface and user experience.

It becomes crucial to differentiate the user experience from the user interface, the user interface is the focal point of the design. The terms user experience and usability should not be confused either, the usability is the quality trait of the User Interface, enveloping if the system is simple to grasp and comprehend, effective and efficient to utilize. This becomes way too important; hence the total user experience becomes a comprehensive notion.

The definitions of user experience and user experience have always been a topic of discussion for a long. These terms did not come into existence at the same time, which makes these topics much more confusing. They did not formulate at the same time. User interface emerged when computing was established in the market.

User Experience was conceived in the 90s by Don Norman when he worked for Apple. He explained in an interview that user experience revolves around the experiences of the users that they have with the product, the company, and the services provided by it.

User experience and user experience are disconnected terms, yet they depend on each other. The purpose of user experience design is to focus on the functionality, and it needs to be foremost as it is the physical structure on which the whole user experience revolves. But the user interface design comes secondary as it comes after the user experience gets completed.

Turning towards the key responsibilities of the user experience designers they must be fluent in content or strategy, i.e., customer analysis, competitor analysis, product structure, and strategy. Secondly, they should be experts in prototyping and wireframing, testing and iteration, development, planning, coordinating with the developers, or with the designers of User Interface, tracking goals, and integration which eventually comes into analytics and execution.

On the other hand, the responsibilities and skills that user interface designers must possess is their focus must be on Branding and Graphic Development, User Guides or storylines, the analysis of customers, and design research. They must also coordinate and communicate and be responsive and active. They must adapt all device's screen sizes, interactivity, and animation, execution with the developers, and prototyping in a user interface.

Both the terms are crucial for the development and delivery of the products, the experiences of the customer generate revenue. So, they are used as an investment in the services or the products. Customers spent a large part of their income on goods and services, so the products they are seeking must fulfill their needs and satisfaction. And if it fails to do so, the interest of customers shifts real soon to other products.

The theories and concepts of user experience and user interface are binary terms in the market, but to carry forward both the terms alongside that is the recent user experience and user interface concepts is the biggest question. Both the terms are often confused even by the expert designers as they are somewhat identical. The clarity is hidden sometimes. They are mainly dependent upon the understanding of people.

5.4. BRIEF HISTORY OF USER INTERFACE

Every part of the hardware consists of User Interface Design (UID) which supplies an experience that proves to be great for users. A good user interface is essential for the completion of the task and making the experience fun-loving. The user interface present in the iPhone or the windows of a personal computer has been in the process of construction for a long.

The User Interface Design has transformed and developed over the past forty years. Batch computing is the point of emergence for User Interface Design, where the user interface of Batch computers consisted of a punched card or the identical media and except for this operating console, people had no relation with the computers that existed earlier.

The user interfaces were not considered important as the software was built to make the most use of the processor. This transformed with the emergence of CLIs that is the Command-Line Interface. The role of CILs was to lower the latency to seconds rather than days or hours as the UI was the set of 'request and response transactions.'

The other important development of the user interface was when the video display terminals were introduced. The command inputs appeared on the screen, and it was modified reversely, it was a faster process than printing. It reduced the cost of production of the printing press by eliminating the requirement of ink and printing materials.

The other interface is GUI, i.e., Graphical User Interface. The clarity about having the digital user interface and pointing systems equivalent to mouse results in an engrossing user interface. The establishment of the first Graphical User Interface was done by the researchers at Xerox Palo Research Center during the 1970s and overtook the computer graphic innovations to Graphical User Interfaces, which has made the world a better place.

The Xerox Star that was released in the year 1981 was the first system that was considered to be the first fully integrated desktop computer. This resulted in advancements in technologies and the release of machines like Apple Lisa Office System 1 (1983), VisiCorp Vision (1984), Mac Os System 1 (1984). In the coming years, the GUI releases more features like color, higher resolution displays, processing power that works better, yet the design of the graphical user interface remains uniform.

When smartphones came into the market, the UI by the end of the 2000s started to change. So, these repositioning in the computing hardware made designers think and work on the interfaces from the beginning. In 2007, Apple launched the best user interface, a touch screen with multiple features of touch and the features and functionalities distributed as apps.

The design of user interfaces revolved around the functionalities of apps. But it was evident that this tendency influenced the user interfaces of desktop and laptop; for instance, windows 8. It has functionalities that were similar to the modern era smartphones or tablets, but there was the choice of a traditional desktop user interface.

Although when Windows 10 was launched, some of the touch or apps functions were the same but it has integrated with the familiar start menu desktop design from the previous releases. The evolution of the User Interface designs has progressed a lot from the textual command inputs and the traditional operating systems.

Though there are numerous similarities prevail between desktop and mobile from keyboards, mouse cursors, touch and voice inputs. These functions and features became wide spread over the recent years and in the coming of age in addition to the development and advancements like gesture functionality, brain-computer interfaces, and built-up reality.

5.5. CHALLENGES IN USER INTERFACE

According to Simon, the partnership and cooperation between the developers and designers are mandatory, and the users must make the most use of digital facilities. The patterns ought to be a boundary object, but it was hard for the developers and designers having different perspectives to set the boundary object as an artifact that is shared equally.

To decide, that the patterns are the tool of the designer or the developer gets tough. They have a different mindset of a pattern library when they explore. To cover up the gap between designers and developers, what is needed is the relationship and collaboration between them, they must listen and work upon each other's ideas and viewpoints. This will bring them together and result in better product and pattern designs, and if they do not come up and work together, it becomes the biggest challenge in pattern designs. According to Wania and Atwood (2009), the essence of pattern libraries lies in the communication between the designers who are involved in working towards the common goal that is the design process. Communication comes first and improvement in designs considered secondary.

So, as this research states, the practitioners focused on improving communication rather than designs, but they did not consider this secondary but they also gave proper attention to the quality used in designs. Their center of attention was to maintain uniformity and consistency in whatever designs they made along with improving the designs, restructure the design patterns and transform into something new and big.

Figure 5.5: Challenges in user interface design.

The practitioners used descriptive patterns in their pattern libraries. And the collection which was stored in the pattern library had various organizing principles. Somewhere it was pattern language, on the other side, it was a pattern catalog.

Pattern language was considered superior to pattern catalog, as it works to the generative pattern idea, yet it was not much appreciated as it brought certain complications and problems to the users. But the pattern catalog was easy in comparison to the pattern language, and hence it proved to be less challenging. The work of designers and developers was to build a framework of rules and assign different roles to the people, to make people aware of what a pattern is.

The secondary challenge that emerged was the decision to arrange and organize principles for patterns. There existed two ways to arrange these principles. The first one was the atomic methodology, which constituted a pattern language, and then secondly a loose collection of patterns in a pattern catalog. It has its pros and cons that there must be a proper decision made regarding the principles for acquiring the goals, objectives, and requirements of the organization.

Thirdly, the pattern library conducts to the adjustments in the workflow, connecting a link between the designers and developers and any change would take time to adjust and adapt for instance a closer collaboration. Another challenge that was faced was accessibility which means that the patterns should be visible to everybody and serve in everyone's interest.

Elaborating these challenges, it is seen that the definition of Patterns is not accurate, and every time is becoming the need of the hour that rules must define what a pattern is and how it works or what should be included and constituted in pattern format. Moreover, the involvement of people becomes necessary, and they should come and work together to set the rules and make final decisions.

They should also take note of whether or not the functionality given, should be made a pattern or not. And in this process, there are numerous roles included, given to people to attain a certain objective of the organization. Rules for patterns are set because no framework of rules results in the emergence of several patterns.

To set the structure of rules, it is mandatory to choose the right people who are fit for the job, by conducting interviews, with the help of a pattern catalog, the users face fewer complications, as the patterns are set accordingly, towards the more inherent causes rather than an unpredictable hierarchy of layers.

These set hierarchies of a layer that are used and stated earlier are a step ahead towards the generative patterns as the clarity between the connection of patterns are visible. The proper organizing principles cater to the needs of users of the pattern library and serves in the interest of the organization by attaining the long-run goals of the organization.

While talking about accessibility as a challenge, it has been seen that there exist different understandings of patterns in the pattern library, different roles exist through which it is evident that the authentication for patterns must remain distinct.

The pattern library must be a shared and common artifact, designers and developers having different roles and needs. The interest of designers must lie in the use cases, while the developers must be interested in the sample code. And, to maintain a balance among both of them, different views of the pattern library must be given to the developers, designers, and others who are involved in the process of designs.

The other problem involved inaccessibility is the reusability of patterns. Patterns, as stated earlier, are the demonstrated solution for the problems that recur as stated by Alexander, 1779. People either made their libraries of patterns or they created their patterns. Although, to make the use of patterns in the truest form in an organization, the patterns must be kept in one place.

However, when we look up to the design problem collaboration, it is considered as a challenge because design problems are considered as 'wicked,' there is no accurate definition of it that exists. They are not monotonous, they keep on changing, and it needs the involvement of a lot of stakeholders.

Due to the absence of a shared common language, there is an inconsiderate amount of difficulty as people came from different backgrounds. Due to this lack of communication, the exchange of ideas and opinions is not shared properly. Patterns are seen through the gap between people from various realms.

The main aspect of the pattern is that they represent the 'proven design knowledge' and they do so better than guidelines. Patterns or the libraries of patterns do not come without challenges; they are aligned to problems and become usable for the designers when these problems and challenges are overcome by them.

These bodies of patterns when once clear and accepted, people could work better towards a real pattern language for User Interface designers. A well-explained and defined format is required, and also a focus on user interface design patterns is mandatory as it contributes to the progression and development of the body and library of patterns.

5.6. WHY DESIGN PATTERNS ARE SUCH POWERFUL DESIGN AIDS

Websites and apps generally have an ordinary appearance and carry a typical theme mostly because of design patterns such as global navigation and tab bars. The UI design offers a quick fix to this problem as it offers the usage of design patterns as a tool to build interfaces that solve a problem—for example, a date picker design pattern allows users to swiftly choose a date in a form.

Therefore, user interface design can perform the role of design blueprints that authorize designers to select and hand-pick the finest and frequently used interfaces for the particular context the user is faced with.

Figure 5.6: Design patterns provide assistance in powerful designing.

Ordinarily, each pattern contains the following:

- A user's usability-related problem.
- The context/situation where that problem occurs.
- The principle involved in the error – for example, error management.
- The definite solution for the designer to execute to tackle the fundamental problem.
- Why – the reason as to why exactly the pattern occurs and how it impacts its reliability and useableness.
- Examples – which highlight the pattern's successful real-time implementation (e.g., screenshots and descriptions).
- Implementation – certain patterns incorporate comprehensive directions.

5.6.1. Common User Interface Design Patterns

Listed below are some of the most familiar User Interface design patterns:

(i) **Breadcrumbs** – Utilize linked labels that dispense secondary navigation that precede from the front to the current site page in the hierarchy.

(ii) **LazyRegistration** – Forms can turn away users from registration. This particular sign-up pattern allows users to sample all the products that a site/app offers without any charges and acquaints them with the feature. Later, a sign-up form pops up on the screen. For instance, Amazon permits all users unlimited and open navigation and also an unblocked cart-loading feature before it prompts the users to sign up.

In situations wherein the content can only be accessed by registered users, or in cases where general users browsing the website/app are asked to fill in details, low-effort/simplified forms with basic details must be offered.

(iii) **Minimize/Avoid Optional Information Fields** – To avoid confusion, the Required Field Markers pattern must be used to direct users to enter the needed data.

(iv) **ForgivingFormat** – It offers the users to type in data in various formats (e.g., city/town/village or zip code).

(v) **ClearPrimaryActions** – Highlight buttons by the use of color to inform the users of the needed actions to be undertaken (e.g.,

"Submit"). It involves the selection of those actions that are to be undertaken on a priority basis.

(vi) **ProgressiveDisclosure** – It helps to showcase only those features that are pertinent to the task at hand, one per screen. It also offers the special feature that allows segregation of input demands into sections, which in turn helps reduce cognitive load (e.g., "Show More").

(vii) **HoverControls** – This feature covers up all nonessential information on detailed pages to help users effortlessly view only relevant information.

(viii) **Steps Left** – This feature is frequently integrated with a wizard pattern by designers. It displays the number of steps a user has to take to complete a task. Gamification (an incentivizing design pattern) can be applied here to enhance engagement.

(ix) **Subscription Plans** – It furnishes users with an options menu (including "Sign-up" buttons) for membership at certain fixed rates.

(x) **Leader board** – This social media pattern is helpful to enhance engagement.

(xi) **Dark Patterns** – A popular pattern amongst designers; it is frequently applied in e-commerce websites and apps to lead or trick users into performing certain actions. Generally, by its applications, users tend to pay out more than they were looking to and at times concede personal details too. Dark patterns range in harmfulness.

A few examples of this pattern are an unchecked opt-out box as a default to secure customer information, slipping items into shopping carts. Since it ventures into a security breach, only ethical apps and websites must make use of it or those who have empathy with users.

Dark patterns need to be used with utmost caution since they are risky for both users and the website/app. User suspicion or mistrust and negative feedback are capable of irreversible damage to any established brand name.

5.6.2. Taking Care with Design Patterns

- Freely available, user interface design patterns let you save time and money since it allows to copy and adapt them into other design – instead of reinventing the wheel for every new interface.

They also facilitate faster prototyping and user familiarity. Nevertheless, they need to be used cautiously. The wrong choices can prove costly – for example, in the case of:

- Excessive dependence on patterns leads to a poor approaches adopted towards problems and their rectification.
- Fine-tuning patterns to specific contexts.
- To not customize a distinct brand image (e.g., your website ultimately resembles Facebook).
- Overlook management requirements. In case a website creates its specific patterns, it is necessary to furnish information regarding its usage, its problems, and version control and also store them for team access.

On the whole, it is expected that users be provided with familiar frameworks that heighten comfort and satisfaction for the users and evade confusion while they interact with the brand at their disposal.

5.7. WHAT ARE USER EXPERIENCE DESIGN PATTERNS?

UX design patterns are repeatable or reproducible quick fixes to intermittent problems in design. While building and designing a new website, or the wireframes for that particular site, the most critical element to be kept in mind is to churn out the simplest way for users to access and interact with the website.

Figure 5.7: User experience design patterns.

The most common prerequisites that are required of the users are: to sign up, sign in, log in and log out. Almost all internet users gain access to the innumerable web and mobile applications regularly and almost all of these users find it a cumbersome task to re-enter the mandatory information required, i.e., name, email, password, and other information. To avoid this oft-repeated situation, a design pattern can be used to help avoid this unnecessary repetitive action.

The problem: Almost all users face a time crunch and lack the patience to undertake this routine and monotonous task repeatedly to access their information or to finish their work.

The Solution: Find a solution to terminate or considerably reduce the burden of this activity for them.

While each designer could offer their solution to this problem, plenty of practical and convenient solutions have already been designed and proven to be effective. To illustrate that point further one can often see that today most of the websites allow users to sign in by linking their social media accounts while many others do not require a mandatory sign-in until deemed necessary and others save their users' data and pre-populate the sign-in fields (with the permission of their users).

5.7.1. Are User Experience Patterns the Same as User Interface Patterns?

Today, there are numerous job titles in User Experience. The sole reason behind this is the fact that this field originated only about 100 years ago. The field itself didn't come into existence and was relatively unknown exist until roughly the 1990s when Don Norman all but invented it, and conferred himself the title of User Experience Architect. To date, the field lacks complete standardization; some of the job titles and terminology have yet to be standardized, and certain terms can be a tad confusing.

As the name itself suggests, a pattern library is a collection of information, resources, and examples of design patterns usually "housed" on a server and accessed through a website, i.e., websites with examples of problem solutions.

5.8. WHAT ARE USER EXPERIENCE AND USER INTERFACE PATTERN LIBRARIES?

It is often propounded that the two terms (and hence their respective libraries) would house distinct patterns. Irrespective of that, it is often observed that both the terms are often used correspondently. For example, the aforesaid site user interface Patterns shows ideas or solutions to common design shortcomings, such as how to handle empty states that resemble mini user flows.

Contrastingly, Pattern Tap by Zurb would be considered more of a user interface pattern library as it showcases ideas for creating engaging user interfaces rather than user flows. Similarly, on another site, also called user interface Patterns, problems and solutions can be found. As observed, even though both of the above-mentioned sites store what are considered user experience patterns, neither of the two is called user experience patterns.

Note: Google's Material Design guidelines and Apple's OS X Human Interface Guidelines (along with their Android and iOS versions) are in reality-style guides, not pattern libraries.

5.8.1. Why Are UX Patterns Valuable?

It is a natural tendency of people to search for patterns. A fine example of this tendency is that while browsing a web or mobile site, users generally experience an unconscious control over it; are aware of where to click. This is because of a repetitive experience. Yet, sometimes users get it right and at other times are confused and searching for guidance.

When sites integrate relevant patterns and put them to use, users experience satisfactory browsing and have a relatively easy and joyful web experience. While certain experts are varying of excessive use of design patterns since it would lead to all sites look similar.

Nonetheless, the majority of designers strongly vouch for design patterns and agree that they offer remarkable user benefits to following previously established patterns…and then tweaking them a tiny bit when benefit can be derived for users.

According to designer Peter Zalman, in his article Design Patterns, "the UI object or screen intuitively implies its use." Using standard design patterns reduces cognitive strain. If people are already familiar with how objects on an interface will act, it saves them from thinking about what to do next.

Reducing the need for people to think too much is important if you'd like to keep them as users. Salman says that if you can't find an existing pattern for interaction, then you must "be prepared to explain and rationalize, why and how the benefits of developing this custom pattern outweigh the cons."

Design patterns benefit both users and designers alike. Patterns spare designers from having to constantly reinvent the wheel. It is a well-known fact among engineers. They now possess large repositories of open-source code like Github that can be accessed by fellow engineers or can be recycled by them.

Figure 5.8: User experience patterns play an important role in designing a user interface.

It is a known fact that rewriting similar code from scratch takes a lot longer than copying, pasting, and tweaking. Similarly, designers who use design patterns can use make use of spare time in creating and tweaking.

5.8.2. What Are Dark Patterns in User Experience Design?

According to the website by the same name, "a dark pattern is a user interface that has been carefully crafted to trick users into doing things, such as buying insurance with their purchase or signing up for recurring bills."

This pattern library is curated and organized by four user experience designers. The purpose of their collaboration was to document and make public common ethically questionable user interfaces both to generate awareness amongst the public and with the expectation to plead to the companies engaging in this malpractice to shun them.

They disclose some of the purposely poorly designed techniques companies use like forced disclosure, friend spam, and sneak into a basket to take advantage of users. Certain websites intentionally do not permit users to unsubscribe. They call those Roach Motels.

5.9. CONCLUSION

In the conclusion of the chapter, this chapter discussed the basic concept of user interface design. This chapter also discussed the patterns that have been used for user interface design. In this chapter, the difference between a user interface and user experience has been discussed. This chapter also discussed the brief history of the user interface as well as the challenges that have been faced in the user interface design.

Towards the end of the chapter, the importance of design patterns, and why these design patterns are such powerful design aids, which include common user interface design patterns, and taking care with the design patterns have been discussed. This chapter discussed the basic significance of the user experience design patterns. In this chapter, are user experience patterns the same as the user interface patterns that have also been discussed? This chapter discussed the importance of user experience patterns, as well as the significance of dark patterns in user experience design.

REFERENCES

1. Emil, L., (2021). *The Difference Between UX and UI Design – A Beginner's Guide*. [online] Careerfoundry.com. Available at: https://careerfoundry.com/en/blog/ux-design/the-difference-between-ux-and-ui-design-a-laymans-guide/ (accessed on 6 July 2021).
2. Jacob, G., (2009). *40+ Helpful Resources on User Interface Design Patterns — Smashing Magazine*. [online] Smashing Magazine. Available at: https://www.smashingmagazine.com/2009/06/40-helpful-resources-on-user-interface-design-patterns/ (accessed on 6 July 2021).
3. Maria De, L. R., (2020). *What is User Interface (UI) Design Patterns? Everything You Need to Know*. [online] Careerfoundry.com. Available at: https://careerfoundry.com/en/blog/ui-design/user-interface-patterns/ (accessed on 6 July 2021).
4. Martijn, V. W., Gerrit, C. V. D. V., & Anton, E., (n.d). *Patterns as Tools for User Interface Design* (pp. 1–12). [e-Book] Amsterdam, Holland: Tools for working with guidelines. Available at: https://citeseerx.ist.psu.edu/viewdoc/download?doi=10.1.1.35.1182&rep=rep1&type=pdf (accessed on 6 July 2021).
5. Meike, W., (n.d). *Patterns as a Tool for Collaboration* (pp. 1–27). [e-Book] Human-Computer Interaction and social media. Available at: https://www.diva-portal.org/smash/get/diva2:919672/FULLTEXT01.pdf (accessed on 6 July 2021).
6. Noirhomme – Fraiture, M., Charrière, C., & Vanderdonckt, J., (2001). A laboratory of ergonomic analyses for children suffering from cerebral palsy. *Tools for Working with Guidelines* (pp. 35–49). [online]. Available at: https://link.springer.com/chapter/10.1007/978–1-4471–0279–3_30#:~:text=Patterns%20have%20emerged%20as%20a,in%20using%20the%20design%20knowledge.&text=A%20structure%20for%20user%20interface,is%20illustrated%20with%20an%20example (accessed on 6 July 2021).
7. Pete, M., (2020). *UI Patterns for Navigation That Makes Good UX Sense.* [online] Usabilitygeek.com. Available at: https://usabilitygeek.com/ui-patterns-for-navigation-good-ux/ (accessed on 6 July 2021).
8. The Interaction Design Foundation, (n.d). *What are User Interface (UI) Design Patterns?* [online] Available at: https://www.interaction-design.org/literature/topics/ui-design-patterns#:~:text=User%20interface%20

(UI)%20design%20patterns%20are%20reusable%2Frecurring%20 components,the%20specific%20context%20of%20use (accessed on 6 July 2021).

9. Van, W. M., Van, D. V. G., & Eliëns, A., (2001). Patterns as tools for user interface design. *Tools for Working with Guidelines* (pp. 313–324) [online]. Available at: https://www.researchgate.net/publication/2638505_Patterns_as_Tools_for_User_Interface_Design (accessed on 6 July 2021).
10. Vanderdonckt, J., & Ferenc, C., (2001). *Tools for Working with Guidelines*. London: Springer London.

CHAPTER 6

LANGUAGE AND COMMUNICATION: USER INTERFACE DESIGN

CONTENTS

In the chapter, language, and communication in user interface design, several design characteristics for language are discussed. It also shed some light on the user experience and communication with visuals. It also discussed the need of using the right language in conversational interface design. It also explains how to build and maintain a visual language. In addition, it also discussed why one should invest in visual language. In the end, it explains how to build and maintain a visual language.

6.1. INTRODUCTION

It is generally seen that a "dialogue" does not need natural language or even words. Animals engage in persistent interactions that can be considered dialogues. A mime is engaged in dialogues with imaginary or real objects and with the audience.

Two individuals who do not have a shared language can work out a means of communication, possibly as a step to constructing a shared "pidgin language." In this chapter, the main area of focus will be on extended human-computer interactions that are "dialogues" in this overall sense.

All forms of human-computer interaction will be included in this, not just "conversational" interfaces to computers. Consider, for instance, the sequence of events in the removal of a document on a Macintosh.

As one navigates the mouse, the pointer or arrow moves across the display. When the pointer is above the icon that signifies the document, the concerned person presses and holds down the button on the mouse. The icon switches to "reverse video" (interchanging white and black), indicating that one is succeeded in choosing it.

When the pointer, as well as the outline image, extend to the vicinity of an icon in the shape of a trash can, that image instinctively shifts to reverse video, signaling that when a person releases the button on a mouse, the document icon will vanish and effectively be redundant. A person does so, and the sides of the trash can bulge marginally, signifying that the document is inside.

Figure 6.1: Language and communication used in UI design.

The bulging sides indicate that there are current documents in the junk can that can be recovered if needed. This entire sequence can be cogitated to be a dialogue, although there is no use of words. The system, as well as a person, monitors each other; the user communicates by mouse activities and button presses, the system communicates by moving objects, swapping them to reverse video, ensuring them appear and disappear, and transitioning their shape. An interface designer is a designer of interaction languages.

Computer systems are exceptional among artificial devices in letting for a practical, intelligent interaction between person and artifact. The formation of interaction techniques is still in its infancy phase. Several design guidelines are extensively endorsed with diminutive critical inspection, such as "build consistent interfaces."

The discrepancies in natural languages—the naturally happening, continually developing communication media used for day-to-day interaction among people—have been examined, suggesting tradeoffs among competing pressures on dialogue. By conflicting the two domains, by discovering where analogs persist and where they do not, one might get insight into the nature of computer system design and a better-off perspective on the restraints on natural languages.

The "design guidelines" proposed by Slobin (1977) for this rich and multifaceted natural system—human language are tried to discuss here. Languages have not been devised; they have developed over thousands of years subject to many competing pressures, including cultural, political, and religious factors.

However, a natural system such as language has much in common with artificially designed computer systems. Many of the necessities are identical.

Each must act as a communication medium to convey intentions, results, and actions among the participants, each must be learnable by novices, yet effectual for skillful performers. The potential of naturally progressing systems like language to deal with these opposing pressures can be enlightening for the design of computer systems.

6.2. FOUR DESIGN CHARACTERISTICS FOR LANGUAGE

Slobin (1977) has analyzed what we might call the "design characteristics" of language, aspects of the usability and functionality of language that led toward language development and change, and that affects the ease of acquisition by children. He identifies four constraints on language:

- Language should be clear;
- Language should be quick and easy;
- Language should be expressive;
- Language should be processible.

Slobin examined the application of each of these rules both to natural language and to human-computer interaction.

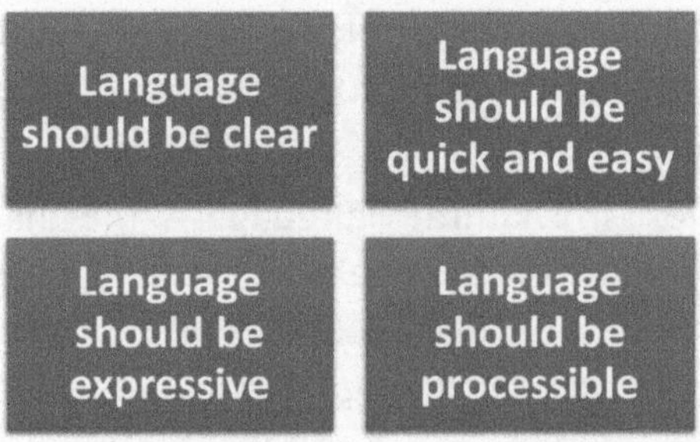

Figure 6.2: Four design characteristics for language.

6.2.1. Language Should Be Clear

6.2.1.1. Natural Language

Clarity can be defined as a consistent "one-to-one mapping between underlying surface forms and semantic structures." Thus, Slobin's concept of clarity resembles consistency as it is usually applied in human-computer interaction.

Uniformity in a language aids learning, both in adults as well as children. Children not only learn more rapidly where it is found, but they impose constancy by discounting alternative constructions (using "I will" or "I will not" whereas on the other side, the adults would say "I'll" or "I won't") or by using a steady form even where it is considered to be ungrammatical (using "hitted" instead of "hit" for the past tense). All-natural languages have discrepancies, the indiscretion of verbs being one well-recognized example.

These indiscretions cause the language learner great challenge because defilements of steadiness mean that a single rule is no longer applicable to a wide class of instances, and rather, many cases have to be learned independently.

Although people have formed more reliable, artificial languages (e.g., Esperanto), none of the thousands of known naturally-forming languages must be wholly reliable. If constancy were as primary a design rule as few have contended, one might have anticipated finding a better presence of consistency in natural languages.

6.2.1.2. Human-Computer Interaction

Computer systems can accumulate similar benefits as natural language systems from a clear, reliable mapping between underlying semantic structures (or actions) and surface forms (or system output and commands).

Here too, steadiness has been shown to assist learning (e.g., Polson, 1988). Although, despite heavy rhetoric advocating steady design and its eminent place in the standard guidebooks, consistency is usually infringed. This is not merely because of oversight—in the best of systems, this violation can advance performance (Grudin, 1989).

One important point of this chapter is to demonstrate that some of the same pressures that influence consistency and stress on clarity in computer systems are found in natural language as well, where they are lucidly viewed to offer significant purposes.

6.2.2. Language Should Be Quick and Easy

6.2.2.1. Natural Language

A language principle that usually conflicts with clarity and consistency is the wish to be easy and quick. This propensity shows up in several ways. The most common words are monosyllabic and short, even in languages that delight long words, such as German.

Language is further abridged through abbreviation or other shortening techniques, attaining efficacy at the expense of regularity, learnability, or even clarity. Irregular verbs and plural nouns are usually shorter than their regular counterpart would be—the discrepancy is presented in the service of efficiency.

Often, as a word intensifies in the rate of use, it is given an abbreviated form: "television" becomes "TV," "automobile" becomes "auto," "picture element" becomes "pixel." Pronouns abbreviate utterances, but at the cost of commencing reducing clarity, ambiguity.

Entire phrases may be eradicated in the cause of efficacy. Although such utterances can mechanically be vague, typically, when understood in context, they are not. It is important to note that the irregular constructions that shorten and simplify will work only if everybody has conversed with them.

Thus, irregularity is most often found with regularly arising constructions—it is the most often ensuing verbs that tend to be uneven. Irregularities pose challenges at the time of the learning process, but once it is learned, they abridge the language process, making the constructions quick as well as easy to use. As long as the irregularities are regularly encountered, they stay learned.

Thus, it is generally seen that the mature native speaker rarely has problems with irregularities: It is only the novice or the learner user who has trouble. An interesting validation of the relationship between irregular language forms and frequency of usage arises as language changes and words transform in their rate of usage.

When the rate of usage of an irregular verb falls, the verb also plunges its irregularities and reverts to a regular form (Bybee, 1988). Therefore, speakers are not loaded with the job of keeping track of language exemptions that occur seldom.

6.2.2.2. Human-Computer Interaction

Does it is possible to find the same push toward non-standard, abbreviated structures in computer interactions? Yes, a recurrent user's desire for easy and quick means to carry out actions results in abbreviation, simplification, and thus, inconsistency. It is generally seen that several computer systems permit their users to generate short keystroke sequences as alternates for longer command sequences:

Some systems even offer these "shortcuts" as typical features: aliases, shell commands, scripts, dedicated function keys, macros, "power-keys or option-key equivalences." Much as the shorter formations in natural language tend to be those that are used with the greater occurrence, shortcuts in computer systems are primarily used for high-frequency operations.

In the Macintosh computer, users often look for a quick way to emit a diskette from the drive and to permitted the memory that the system had kept for it. Originally, two operations were needed: an "eject" command and the action of moving its residual, "greyed-out" icon into the trash can. In a triumph of usage over steadiness, a creative programmer joined these into one operation, conveyed out by moving the diskette icon to the trash can icon.

The operation desecrated notions of several people of reliability and puzzled first-time users, but because of its overwhelming effectiveness, it became accepted widely.

Computer users that form their shortcuts often produce name sets that are effectual, but so unpredictable that they consequently forget the names that they planned (Grudin & Barnard, 1985).

They may miscalculate the occurrence with which these terms will be appraised. Other users, of course, are probable to find these personal shortcuts to be inconceivable. Natural language handles the equivalent problem through numerous mechanisms.

6.2.3. Language Should Be Expressive

6.2.3.1. Natural Language

Natural languages must have influencing expressive ability "... to communicate efficiently, engagingly, suitably, and so forth. The speaker must be able to express the listener's attention, to take into consideration expectations or knowledge" (Slobin, 1977, p. 187).

The critical aspect here is that language must perform in a wide range of contexts, needing a versatility that usually comes into conflict with the other restrictions. To be both efficient as well as expressive, language must be crushed—thus sacrificing a well-defined, consistent mapping between function and form.

Slobin writes, "it is the charge to be expressive which presents much of the intricacy into language." Miller (1951) observed that "the social pressure for a common vocabulary and the convenience of monosyllabic words tend to confine the range of our responses, whereas the effort to distinguish between similar statements magnifies the vocabulary and resulted in the occasional use of polysyllabic words."

This captures the opposing pressures of Slobin's maxims "be quick and easy" and "be expressive." The several tensions push the solutions in opposing ways.

6.2.3.2. Human-Computer Interaction

It is generally seen that the range of expression is thinner in computer interaction in comparison to the language, but as applications mature, the demands for a wider range of articulateness grow, and we find analogous sources of the discrepancy.

Information retrieval systems offer a wide range of search potentials, whereas, on the other side, a simple string search is adequate for a word processor; a professional typographer needs a degree of layout precision not required for the majority of the document preparation. The result is usually unreliable interaction languages of varying intricacy.

Overall, large applications may have hundreds of commands to satisfy the necessities of thousands of distinctive users, who often needs entirely distinctive system performance. One can expect this issue to be of increasing significance as computer systems become more powerful and richer.

6.2.4. Language Should Be Processible

6.2.4.1. Natural Language

The degree up to which the speaker, as well as the listener, can precisely encode and decode language utterances must be comparable. If they ensued at rates that were too discrepant or led to too much error, there would be a sacrifice in communication.

This is particularly challenging in spoken language, because of the non-persistence of sound—the listener has a restricted potential to review what has been vocal, and thus must process it in "real-time."

6.2.4.2. Human-Computer Interaction

Computer communication is persistent: The computer can sustain a record of input and can offer consistent output utilizing a static visual display or by letting ready recurrence of an otherwise visual signal or transient auditory. Even so, a typical restraint to be humanly processible operates in the visual medium as well as the acoustic.

For instance, in the design of visual icons, the comparative size distinction of a trash can and a document in the real world is not recorded onto the interface (it would make one icon too small or the other too large, similarly, one may augment the users' potential to differentiate among objects by amplifying differentiations (Hollan Hutchins, Rosenstein, McCandless, and Weitzman, 1987).

Thus, if it is critical for the users of a system that controls an industrial process to differentiate between 200-gallon and 220-gallon boilers, a designer might use icons that vary in size by 50% instead of a precisely-mapped 10%. This disrupts a clear mapping of semantic information onto the surface form, but offer greater human processibility.

6.3. USER INTERFACE

6.3.1. User Experience

User experience (UX) includes everything from the way packaging looks and feels to communication between the product and user. User experience is also created if the packaging is visually good looking if it is easy to get home from the store and how easy it is to unpack. Good design is a combination of smart interaction and good looks.

The designers of user experience create the interaction between the product and the user. The user interface is where this interaction is taking place. The objective of user experience design is to make the user interface feel like a conversation is going on between two human beings.

The communication should not be technical as well as cold, just like between two robots. The language should be intelligent and respectful. It is important to understand the technical language so that it is understandable for users. The objective should be to ensure that while using the product, users get to their goals, getting the user from point A to point B, and finishing the task without frustration.

The process of user experience design includes a lot of user research, prototyping as well as testing the product with users. The use of prototyping and testing is to discover the way the user feels about the user interface. The only way to get into the mind of the user and to design for the user is by working and prototyping with the user.

An important part of communication with the user is targeting the product for a specific user. Communication with the target user is easier when the product is not designed with every possible user in mind (McKay, 2013; Ussai, 2015; Zumbrunnen, 2015; Norman, 2016).

Personality, as well as the brand of the product, is also considered as a big part of communication. Branding means picking the target audience. It is necessary to think about the brand in the design process of the product for picturing the target user and even making up the target users.

The products should not be designed for everyone. In simple words, they should be designed for very specific users.

This helps the communication between the user and the interface and makes it feel more personal for the user. It can become more memorable by giving the product personality and tailoring it for a specific target audience.

This makes the job of the designer easier, not only the user experience design but also the visual design. Things such as the user's education, age, sex, occupation, location, interests as well as task affect the design. In this way, it becomes easier to speak the language of the user (Jensen, 2013; Puri, 2014; Zumbrunnen, 2015; Ussai, 2015). User experience is not just the user interface or the app.

It is everything tied to that and the experience of the user with the product even including the package and the shopping experience. Everything affects the mood of the user, how the user feels about the product together with the surroundings. In case the product is an app, the hardware (in this case a smartphone) used by the user influences the experience.

Apple is well known for its good user experience from sleek hardware, easy packaging, and smart software. The experience of a product is good in case you are picking the product from the store for using it daily. Apple is doing well due to its good user experience design. Apple begins the designing from the experience rather than the technology.

Apple is worth paying a bit extra money as users get a positive feeling from their product. The company knows what the user wants, and this makes it so great. It does not give the user all the freedom as well as the possibility to modify as Android and PC does. It might be possible that some users need the freedom to modify, however, regular users do not know what they need and want.

Designers need to know the user. The user might get confused if limitless possibilities to modify the product are given. The designer should have empathy as well as the skill of putting themselves because of the user. You can understand a problem when you can truly feel the problem (Jobs, 1997; Marsh, 2014; Norman, 2016).

Feedback is an important part of communication between the user and the computer. It is a demonstration of the interaction of the user with the product and helps to know if it was successful or not, and why. Mostly, feedback is shown when something is correct or incorrect, loading or in case a selection is made. It also means that the user interface gives responses to the tithe user at all times.

Responsive interaction gives satisfaction to the user and makes the element feel real when they can see that their actions have an impact. Feedback can be given with sound or vibration, smart graphics, or motion. A very good example of excellent feedback is the mobile game Two dots.

All those dots will explode when you connect the same color dots into a square. The whole screen will fill with color; the phone will give a happy sound and vibrate. This action will also help to get you through the easy level. All this feedback creates a good satisfaction level for the user. Feedback should be given to the user regarding everything related to the task they are doing.

For instance, it should be told somehow in case something is loading. Otherwise, it might seem like the program or the computer crashed. This can make the user frustrate. Problems, as well as errors, should be explained to the user in simple language rather than engineering language.

The error message should have the problem together with the solution to the problem instead of zeroes and ones or another technical language (Jensen, 2013; Ussai, 2015; Zumbrunnen, 2015).

6.3.2. Communication with Visuals

A very important part of mobile user interfaces is visualizing, however; there are cases in which the user interface does not require to be visual. The design is more important than visualizing when the computer is used as a tool. Visually beautiful interfaces are more logical and more pleasant to use when space is designed well.

The user interface has to support the content and bring it out, as it cannot overpower the contents of the interface. First, attention should go to the content, in case the user interface is designed in a good manner, it is something that users will not even notice.

In the year 1988, Norman stated that a good design is harder to spot than a bad one (Norman, 1988, 12). Everybody used realistic-looking graphics or skeuomorphism design when mobile graphics just started coming up. This was overcome through completely flat design just using shapes as well as colors. In addition, later in the year 2014, Google brought up material design 2014. It is also known as quantum paper and uses a flat design however, adds some dimension to it.

Generally, there are flat planes in 3D space. Space the design still looks flat even though the elements are in 3D because it is viewed from the top and the drop shadow adds more dimensions. The material design of Google is a visual language.

It is design principles for typography, color, space, grids, scale, motion, and use of imagery. The inspiration came from real paper, and the objective was to have more liberal use of responsive animations, grid-based layouts, and transitions, padding as well as depth effects such as shadows and lighting (Google Material Design, 2016.)

Motion is a very big part of the material design of Google. It is part of the communication and uses animation principles of Disney. Motion clarifies all the steps as well as commands that the user is taking and assists them to navigate where they are.

Through this spatial awareness, it becomes easier to get into the flow of using the user interface. Motion helps to connect the dots between every action, and as a result, it gets easier for the user to observe the relationship between information. To make navigation easy for the user, the information should also be divided into smaller districts so that it can become easy to manage (Google Material Design, 2016).

Figure 6.3: Communication with the visuals.

The graphics should be very simple to use motion so that the user will not get sidetracked by the entire color, detail, or any other type of abundance.

In case there is an existence of motion communication in a user interface, it should be as minimal as possible as a user interface is only there to support the content so that the user gets the information that they are looking for.

The application of Instagram is one of the good examples of a minimalistic user interface. Instagram published the new updates for overall design and user interface in spring 2016. This new interface gives more importance to the content.

Moreover, an application such as Instagram is all about the content. Before noticing the interface, people should notice the content due to the reason that the interface is there to support the content. The new colorless interface of Instagram does not fight with the entire content of the user such as videos as well as colorful pictures.

There is no need for colors or other extra designs if the user interface is designed well. At present, the interface of Instagram has only colors in its notifications that are pinkish-red and tags and links that are blue. The old interface was comparatively darker and had some shades of colors. Earlier, the interface also took a lot of space, however, at present, the top, as well as the bottom bars, are shorter, giving more space.

6.4. CONVERSATIONAL INTERFACE DESIGN: USING THE RIGHT LANGUAGE

Modern conversational interfaces help users to interact with machines in a user-friendly way with the usage of written or spoken words. Messaging platforms with billions of users to be reached will become more powerful and only multiply with the passage of time. A natural human conversation can be stimulated by the advancement of voice UI technology and machine learning that will further transform business communication.

Although, the use of voice interfaces is the latest trend of the future, however, most conversational interfaces are still based on text. The creation of the right conversational UI and chatbot text takes some creative writing or film script, technical writing, transcription, and interpreting, together with a passion for good UX copy and insight into how people write and speak.

The international objective is to model the user experience after a natural human conversation with inherent rules of cooperation and clear logic. In simple words, anyone can contribute to chatbot building through the usage of the language skills that lie at the heart of conversational interface design.

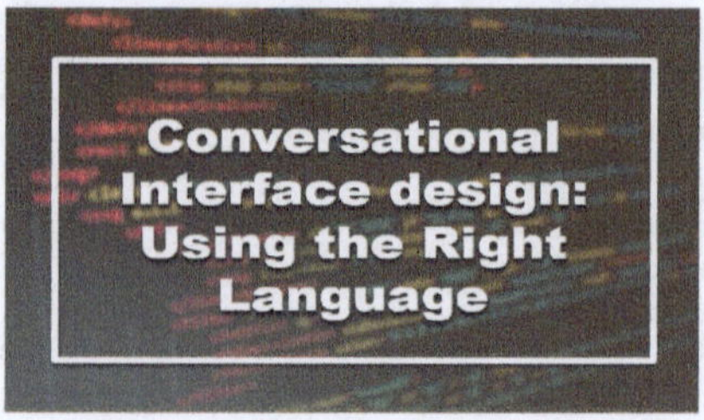

Figure 6.4: Conversational interface design: using the right language.

Continuing the series of blog posts on chatbots, we would like to highlight the importance of language design for chatbots as well as a few relevant lessons that we have learned at the time of creating storylines for the prompts, flows as well as messages to display. To delight the users, these practical tips will help you build and design a useful chatbot.

6.4.1. Design the Bot to Meet the Users' Needs

Chatbots assist users by increasing convenience, providing expertise, facilitating their interaction with a brand, etc. A bot is designed to make this problem solving as comfortable as possible for them.

The designing the chatbot began with determining the bot's perimeter, i.e., what it should do and what it should not do. A rule of thumb to define the right perimeter of a chatbot is that it should be aware of the meaning of every term it uses.

To manage the expectations of users, an idea about the bot's perimeter should be given at the very beginning of the interaction. Design an initial message, speech bubble, or animation with an appropriate text. It should not need a long time scrolling as well as reading. Hook the users with a question at once and avoid stating the obvious or talking down to the users.

Aim at developing conversational experiences that fulfill the preferences and needs of some users. We should try to guess the way they are most likely to react in particular situations.

Remember that some may require comparatively more support whereas some others may be trying to 'break' the bot by asking tricky or silly questions. Your choice of natural language processing (NLP) should also be tolerant of partial sentences, phrases, and grammatical/spelling typos and errors.

The system must be able to process abbreviations, slang, emoji, jargon, etc., as well as to deal with rude conversation and user frustration. The bot should have multiple variations of 'Please help me out here and 'Sorry, I don't understand 'in store.

Avoid repetitive and lazy responses. The chatbot should use synonyms for those terms, slang equivalents, and abbreviations – all that the customers might use in addition to the domain-specific terminology. "Acknowledgers" should be randomized to make the experience feel more natural and fluid.

It is not possible to design a customized series of responses and questions with the language targeting or adapted for each type of user. However, chatbots that 'remember' the user's past topics, preferences together with defaults can display different content based on collected data.

Consider A/B/C testing with multiple learning and responses as well as find out which ones work best. Machine learning will be most helpful because it learns what does work and what does not work and as a result, it will help you design conversation that flows automatically.

A chatbot is considered in case it can recognize when and where the conversation takes place and provide empathetic contextual responses. The opportunity for a more relevant personalized bot experience can be opened up by the mere use of language with its emotional load. Design the bot training with a single goal in mind: natural language understanding and leave aside your business goals for the time being.

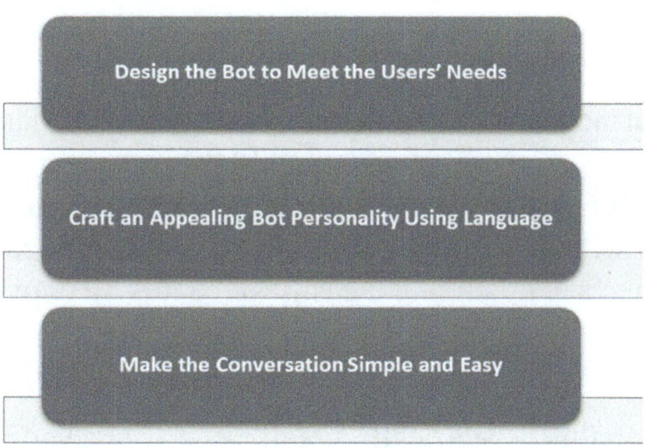

Figure 6.5: Conversational interface design.

A chatbot is considered good if it uses the implicit set of rules that are followed in regular human conversations. Building blocks of a healthy conversation include:

Turn-taking in a conversation relies on signals that are passed by people back and forth. Incentives are given to the user for responding as well as continuing the dialogue.

The flow of the conversation can be tracked with the help of threading through a coherent context and the way each step links to the one before and after. Chatbots should use NLP to tease out the idea behind confusing statements and fragments and pay close attention to conversational threads.

Often people leave some things unspoken however, still can be understood as well as 'read between the lines' in the speech of other people.

A bot should be capable of understanding what the user is asking based on contextual cues as well as use syntax to respond to what is said. The software-based mechanisms should be there to compensate for the illogical and non-mathematical nature of the human language. This is especially important with verbal conversational as well as voice-activated UIs.

In a conversation, user behavior depends on their situational context and expectations of a dialogue based on previous experience. The worst case is misspelled words and apparent mistakes. Conversational UIs must be ready to support a variety of data formats, synonyms, and styles together with several other variations to make sure a frictionless experience.

Text shown on mobile devices must be quick to read, skim as well as use on the go. Generally, it is suggested to keep the bot's messages in the size of a tweet for the sake of convenience. Still, there are many situations where it is considered the best to respond with a long list, text, etc., so that the user can keep the screenshot of some important information.

The last but not the least point is about the 'right language.' One of the practical advantages of conversational UI is localization. The back-end database can be set in such a way as to add in translations to other languages. The translation will be simple and allow launching a localized version of your chatbot interface quickly and easily since it is all copy-driven.

6.4.2. Craft an Appealing Bot Personality Using Language

A certain 'personality' specifically created for your chatbot will give you comparatively more control. For instance, a 'friendly' bot can make the difference between meeting a new friend vs. being surveilled during personal data collection. Natural-sounding and empathetic bots can upsell according to the needs of users. A distinct personality of the bot also assists in reducing the volume of language that will be used during the conversations.

When building a bot for Slack, Telegram, or Facebook Messenger, their graphical UIs limit the visual component of your conversational interface design. You can only make your bot unique by designing its conversations with the visitors. The most favorable thing is that the medium permits to development of whole characters seemingly living behind the messages.

All this begins with getting aware of your users as well as the context, such as their age, expertise, etc. A clear persona of the bot should be visualized 'behind the curtain' that can resonate with the users as well as develop a style together with a tone of voice that fits the persona. The prompts of the chatbot should align with the conversational tone and style. The vocabulary options available for selection should project that persona as well.

While talking to potential clients, a polite chatbot uses 'we' as well as 'our' and says 'you' or 'your' only when it is necessary. It might motivate them if it failed or compliment the user on a successful session. It should not press the user by using uppercase prompts or sound condescending.

The traits of your brands are embedded in your chatbot interface. Attention should be paid towards those personality traits that determine the bot's errors, greetings, small talk, and requests for user feedback. Would it sound condescending or use formal language? Will it compose lengthier sentences or give curt responses?

Can a chatbot use an occasional curse word or sound witty? Unexpected responses, single words, or phrases in the repertoire of your chat box will keep the users engaged throughout long conversations. The user experience can also be enhanced by humor, and if a user explicitly asks the chatbot to tell a joke, it cannot help but comply. In context, just make sure to use humor wisely and respecting the cultural nuances.

The interaction will be more natural, smooth as well as seamless when your bot will be just like a human being. Emotion sensing helps bots powered by artificial intelligence to understand the feelings of the users behind every message and react accordingly.

Another strategy for more humanlike chatbots is to reflect how people converse with each other during online as well as offline chatting. When the user speaks in slang language, the use of the same slang by the bot makes it more relatable and eases confusion. Especially if the bot has a voice, filler words can help conversations sound more natural too.

6.4.3. Make the Conversation Simple and Easy

The chatbot should have simple, clear, and attainable goals so that the user can achieve them without overthinking and typing/clicking too much. Attempt to simplify choices of the user so that they can get what they need quickly. The topics of the conversations must be related to the issues that the chatbot is expected to resolve.

The system must be able to provide simply expected dialogue directly to the users. Elements of graphical UI guide the conversation; however, the language you use is your primary instrument. It enables the users to choose from multiple options quickly.

It includes ways to rephrase queries, clarifying questions as well as prompts for solving errors in your conversation flows.

Make the texts of the buttons or several other widget labels supporting the chatbot prompts action-oriented as well as short. Usage of imperative forms of verbs to clarify to the users what selecting that option will entail. Prompts should begin with a capital letter and end with a punctuation mark, for example, a period or question mark.

One should not forget that users would be primarily viewing your chatbot interface on their mobile devices. Moreover, in most of the messengers, the message bubbles stick to a side of the screen that makes longer messages look like flabby blocks of cramped text.

Try to make the messages of the bot concise as well as short. Taking turns with short utterances, just like a lively conversation, prompts users to act immediately, keeps them engaged, and helps prevent misunderstandings resulting in back-and-forth dialogue.

Preference should be given to short words instead of long ones. In a voice interaction, long as well as complex sentences are considered undesirable. On the other hand, responses that are too short make the conversation sound mechanical, such as a simple 'Yes' or 'No.' Therefore, it is vital to arrange a balance and test different utterances that align with the personality of the bot.

6.5. HOW TO BUILD AND MAINTAIN A VISUAL LANGUAGE

The design provides a platform for communication. We use language as a tool for communication with other people. Writers use words to communicate with their readers, while designers use visual language to communicate with their users.

Fonts, colors, shapes, visual elements such as icons – those are elements of design language. Effective design language streamlines communication. Moreover, it is hard to understate the importance of having a visual language when it comes to selecting the right visual elements for a design.

Visual communication is indeed a complex thing. Many designers attempt to oversimplify the role of a visual language for purely aesthetic purposes. It is a fact that a visual language is a lot more powerful than that. This topic of the chapter will discuss visual language and how we can create one for our next design project.

Figure 6.6: Build and maintain a visual language.

6.5.1. What is a Visual Language?

A system of communication using visual elements is known as a visual language. Visual languages are not only applicable to digital products but are also applicable to everything that can be visualized. Visual languages assist users in comprehend and perceive visible signs.

An example of visual communication is a map because it transmits visible signs (colors, text, shapes) that an observer decodes with the usage of vision.

6.5.2. How a Visual Language is Different from a Design System

The question many designers have in their mind is whether a design system and visual language are the same or different thing. The answer is no, but both the concepts are related.

Usually, design systems contain a style guide or/and component as well as pattern libraries together with a set of instructions that gives the team a clear and simple idea of how to put the components together. Design systems allow the product team to create products and manage the chaos in a scalable way.

The design system constitutes the overall design language and visual language is a part of it. The visual language aims to focus on the visual comprehension of information.

It is not only regarding the way styles/components can be used together to make a consistent UI, but it is also about the reason for choosing the usage of styles/components in the first place. In simple words, the visual language provides an explanation or background as to why things have been put together in a particular way.

6.5.3. The Elements of a Visual Language

A visual language consists of visual units. There are two kinds of visual units and that are atoms and molecules. Color, shape, space, typeface, form, and proportion are all atoms. Atoms can be merged into larger, more complex objects known as molecules. For instance, input field, label as well as button atoms join together to create a search form molecule.

Figure 6.7: The elements of a visual language.

A component can be a combination molecule as well as atom. The top header of the website, which itself would be a component may include the search form.

Visual units have two important facts and that are worth mentioning:

- In a visual language, there are no random visual units. There is intent and meaning in every element of the visual language. It is especially important to define elements of the atomic design in such a way that gives semantic meaning in view of the fact that all other visual units will be based on them.

- No isolated units should be there in visual language. In a visual language, every single unit should be a part of a greater whole.

6.6. WHY YOU SHOULD INVEST IN A VISUAL LANGUAGE?

The global objective of building a robust visual language is to upgrade the experience of the user. Visual design is about using the visual aspect of a product to improve the experience of a user of that product. Therefore, we can say that visual language is a foundation of visual design. The visual language also creates a better UX in three main ways that are mentioned below.

6.6.1. Shared Vision

A single person rarely builds digital products. A few different teams work on creating a product in most cases. When the size of the team is large, it becomes more challenging to create a consistent experience.

Members of the team tend to produce new styles as well as solutions, and this can cause a disjointed user experience without a shared visual language. Visual language can define a more systematic way to leverage team efforts and provide guidance.

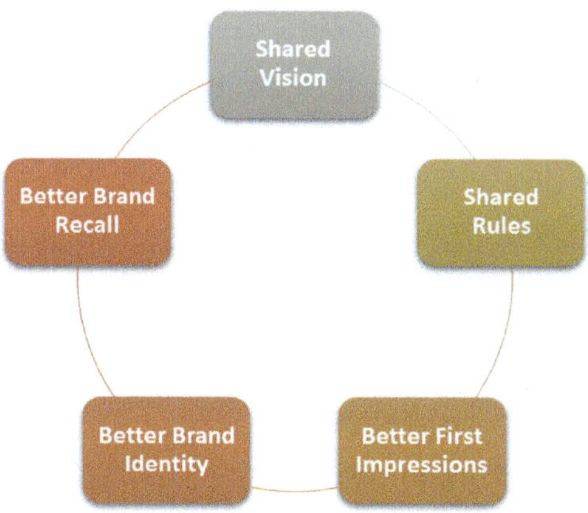

Figure 6.8: Reasons to invest in a visual language.

6.6.2. Shared Rules

A visual language has rules that enable developers as well as designers to communicate with one another, just like any spoken language has rules that allow one human to communicate with another. Hence, it enables your brand to communicate with your users. It is observed that many times rules create constraints for designers. Here, the term 'constraints' might sound like a restriction to creativity, however in real terms; it helps designers to work more efficiently and effectively.

As compared to many other design fields such as industrial design, digital product design has only a few physical constraints. Moreover, it can often lead to situations where designers propose solutions that can cause disjointed user experiences. We create constraints by establishing clear rules that prevent team members from creating bad UX.

6.6.3. Better Brand Identity

A brand provides a platform that helps customers to perceive you. A successful product is unable to survive in the market in the absence of a brand. A visual language is one of the strongest ties that can be used by a designer for effective branding. Particularly, it assists in creating better brand recall as well as better first impressions.

6.6.4. Better First Impressions

Various researches prove that creators of the product have a time of only 50 milliseconds before users make their initial judgments regarding products. Even before users start using your product, the visual design is the first impression your product will make on them. Moreover, it is something that impacts their decision that whether they will become your customers or not.

6.6.5. Better Brand Recall

Users rarely think about the visual language a product uses, but they tend to remember it better when they interact with a product that has a strong visual language. Creating a strong visual identity is a goal that design teams should state when working on visual design. This is the personality of the digital product! The colors, typefaces, photos, illustrations, animations are all part of a brand, and they should be designed in a way that helps people remember the product. When an authentic design language is followed consistently, it creates recognizability for the brand.

6.7. BUILDING AND MAINTAINING A VISUAL LANGUAGE

A great visual language very rarely happens by accident, just like great things. It is usually the outcome of intensive as well as long work. The seven rules mentioned below will help to make the process of creating a visual language more efficient.

6.7.1. Understand the Anatomy of the Product

Even before starting work on a visual language, you need to conduct a UI audit of your client's product to get a clear understanding of what colors, shapes as well as fonts are being used at present. It is also worth getting any guidelines from stakeholders about the brand. These will assist you to understand the general look as well as the feel of the branding of the company.

Figure 6.9: Understand the anatomy of the product.

In the absence of all this understanding and information, you will work completely in the dark and at risk. It will result in a wastage of your time in building something that might be undone. Time should be spent in acquiring the required information and analyzing it. When you begin working on the language, it is important to make sure that your design decisions follow the findings you have learned.

6.7.2. Understand How Users Perceive Your Brand

As mentioned earlier, visual language is considered an integral part of brand identity. Visual language must represent what or who you are. It is always essential to have a clear understanding of who you are designing for while working on visual language.

You should invest enough time in user research to have a clear understanding of your target audience and how they perceive your brand. This understanding will provide you the basis for your visual direction. Your visual identity should be reflected in every bit of your design.

6.7.3. Create a Dictionary for Your Visual Language

Visual language begins with a dictionary that defines visual units as well as their meaning, just like spoken language begins with words and meanings. Style guides together with a pattern library (reusable building blocks which can be grouped and categorized) are the pieces that make up the dictionary for your visual language.

This dictionary should be able to explain the reusable visual units with clear meaning. It can be challenging to map elements of design with clear purposes and meaning, but it is possible to evaluate the meaning of elements by the prism of communication. The template mentioned below can help you identify as well as communicate the meaning of each unit:

The design element used in the library/pattern helps to communicate meaning as well as purpose. This technique should be used whenever you evaluate a specific visual unit. For instance, you could say, "The vibrant orange used in the signup form helps to communicate the importance of this action for users," in case you are selecting a color for a primary call to action button on a landing page.

6.7.4. Define Clear Design Principles

Establishing the clear syntax together with the semantics of the language is the next step after defining a dictionary. The grammar of the language is design principles. They define rules on how to use a specific element at the time of creating a design. Rules of design principles should always be kept in mind at the time of designing the product.

Design principles should possess the following characteristics:

(i) **Genuine.** They should reflect the philosophy of product design that should be followed by your team.

(ii) **Actionable.** Design principles should not include general suggestions on the ways of designing products. For instance, they should not use words like, "Make things beautiful." As we know that no one wants to intentionally build ugly products; therefore, it does not mean anything to the person designing the product.

(iii) **Clear.** Design principles should be clear and simple so that everyone who is participating in product development can understand them.

(iv) **Inspiring.** Design principles should inspire creativity. They should provide enough room for exploring creative design solutions and guide the work of the team in a unified direction.

For instance, medium is a popular blogging platform and one of its design principles is direction over choice. The Medium team describes the same in the following way: "This principle was often referred to us at the time of designing the Medium editor. We purposely traded type, color as well as layout choices for direction and guidance. The direction was more appropriate for the product due to the reason that we wanted people to focus on writing instead of getting distracted by choice."

It is a clear directive to build something actionable as well as authentic for the Medium designers. To learn more about design principles, I highly recommend reading the book Design Systems by Alla Kholmatova.

6.7.5. Document the Process

You should never overlook the documentation process no matter how tight your deadlines for creating a visual language are. Comprehensive documentation regarding the visual communication system is something that can be extremely necessary for the design process of a product.

A common cause of confusion at the time of the implementation phase of a design is the lack of thorough documentation. Throughout the creation process, creating documentation allows for smoother decision-making as you will have a better rationale behind each decision.

6.7.6. Stick to the Rules You Have Established

It is vital to stick to a set of guidelines for a visual language once a product team has developed and reached an agreement on them. Inconsistency is the biggest misstep while building a visual language. Inconsistency can be seen when team members do not follow guidelines. We should not forget that there is no use for a language if nobody wants to speak it.

6.7.7. Think of the Visual Language as a Living Organism

Languages that can be spoken tend to change over time, as cultural influences shape as well as impact them. Visual languages are very similar to them. Hence, we can say that a visual language should not be a set of rigid rules but instead should be an evolving ecosystem that grows together with a product. Moreover, this ecosystem should be adaptable to changes easily.

6.8. CONCLUSION

In the end, it is concluded that language and communication are some of the most important elements in the user interface design. It is highly imperative to convey the message clearly, efficiently, and effectively to the user who comes to our website. The user interface should always be interactive by incorporating audio and visual features in websites. It is always important to convey the right language by keeping in mind the target audience who explore the website. For instance, the culture and tradition of one country are different from another, so this should be kept in mind while setting the website for one country in comparison to another.

REFERENCES

1. Agnes, K., (2021). *Language and Communication: Essential Concepts for User Interface and Documentation Design – Open Research Online*. [online] Oro.open.ac.uk. Available at: http://oro.open.ac.uk/27398/ (accessed on 6 July 2021).
2. Andreev, R., (2001). A linguistic approach to user interface design. *Interacting with Computers, 13*(5), 581–599.
3. Grudin, J., & Norman, D. A., (2021). *Language Evolution and Human-Computer Interaction*. [e-Book] Available at: https://www.microsoft.com/en-us/research/wp-content/uploads/2017/01/CogSciNorman1991.pdf (accessed on 6 July 2021).
4. Kemilä, J., (2021). *Communication With Motion in User Interface*. [e-Book] Available at: https://www.theseus.fi/bitstream/handle/10024/116132/Kemila_Jenna_Communication_with_motion_in_ui.pdf?sequence=1&isAllowed=y (accessed on 6 July 2021).
5. Kukulska-Hulme, A., (1999). *Language and Communication: Essential Concepts for User Interface and Documentation Design*. Oxford University Press.
6. Six, J. M., (2021). *Choosing the Language for a User Interface: UXmatters*. [online] Uxmatters.com. Available at: https://www.uxmatters.com/mt/archives/2008/11/choosing-the-language-for-a-user-interface.php (accessed on 6 July 2021).
7. Marcus, A., (2021). *Principles of Effective Visual Communication for Graphical User Interface Design*. [e-Book] Available at: https://www.diliaranasirova.com/assets/PSYC579/pdfs/05.2-Marcus.pdf (accessed on 6 July 2021).
8. Nielsen, J., (2021). *10 Usability Heuristics for User Interface Design*. [online] Nielsen Norman Group. Available at: https://www.nngroup.com/articles/ten-usability-heuristics/ (accessed on 6 July 2021).
9. Redish, G., & Barnum, C., (2019). Overlap, influence, intertwining: the interplay of UX and technical communication. *JUS*. [online] Uxpajournal.org. Available at: https://uxpajournal.org/overlap-influence-intertwining-the-interplay-of-ux-and-technical-communication/ (accessed on 6 July 2021).
10. Renk, R., (2021). *Plain Language for UX Writing*. [online] Medium. Available at: https://uxdesign.cc/use-plain-language-in-ux-writing-d7d5b0ea35f1 (accessed on 6 July 2021).

11. Shopify, (2021). *How to Build and Maintain a Visual Language.* [online] Available at: https://www.shopify.in/partners/blog/visual-language (accessed on 6 July 2021).
12. Slesar, M., (2021). *Conversational Interface Design: Using the Right Language.* [online] Onix-systems.com. Available at: https://onix-systems.com/blog/conversational-interface-design-using-the-right-language (accessed on 6 July 2021).
13. Studio by UXPin, (2021). *What Actually Constitutes Design Language?* | UXPin. [online] Available at: https://www.uxpin.com/studio/blog/design-language/ (accessed on 6 July 2021).
14. The Interaction Design Foundation, (2019). *The Use of Language and the User Experience.* [online] Available at: https://www.interaction-design.org/literature/article/the-use-of-language-and-the-user-experience (accessed on 6 July 2021).
15. Vukovic, P., (2021). *7 Unbreakable Laws of User Interface Design.* [online] 99designs. Available at: https://99designs.com/blog/tips/7-unbreakable-laws-of-user-interface-design/ (accessed on 6 July 2021).

CHAPTER 7

CHALLENGES AND OPPORTUNITIES IN USER INTERFACE DESIGN

CONTENTS

This chapter explains the various challenges and opportunities that have been addressed in user interface design. This chapter also explains the basics of user interface design, such as choosing interface elements, and best practices for designing an interface. This chapter also sheds light on the several numbers of challenges that user interface designers face and how to over those challenges.

This chapter also addresses the biggest challenges that have been faced in the process of designing such as working within fast turnarounds, adapting to technological change, cultivating human experience, and so on. This chapter mentions various common issues that have been coming across in the UX process, along with the solutions to those issues.

7.1. INTRODUCTION

From the year 1999 to 2018, this paper presents an important analysis of a literature review, which has the findings in design issues. So, in this study, there is a discussion on the study about all the issues related to design and user interface, and also gives the solutions to make the designs or user interface more attractive and understandable.

To solve the main issues of user interface, this study is the guideline. For modern applications, it is important to secure the system. Over the years, the use of the internet is quickly growing. In this way, they let the user attach with systems from everywhere because of this fast-traveling lifestyle.

In the system, when a user is ignoring the functionalities, then the system is not secure but, in other systems, there exist some threats that can harm the system. For example: To use a device such a system is not secure for such a user, when the user is not educated or has no knowledge.

Figure 7.1: Challenges and opportunities in user interface design.

Through an audit of existing writing, this chapter points judgmentally, as the significance of exchange off or adjust amongst convenience and the security of the framework. By utilizing the heuristics assessment, intellectual walkthrough, formal ease of use review, pluralistic walkthrough, highlights examination, consistency investigation, standard review, and numerous more techniques the UI is being checked or estimated for all kinds of items.

For all the current telephones the creator played out the study to look at the ease of use. Two models are being discovered as the outcome for the simplicity of the client, client fulfillment and learnability. In the view of a proposed structure, the survey results that UI is easier to understand than the broad-based UI.

About the human-computer interaction, this is article is all about and along with it is also about the user interface design issues in the HCI. For all the users, the main issue in the design is that the interface is not understandable, so this is the reason when a person does not understand the interface, then he is not able to use such software.

When the user and the computer both can interact with each other, then it is known as the user interface. For example, in using the input devices, the computer and user both are attracted to each other. The main types of the user interface are as following:

- Command-line;
- Graphical user interface (GUI);
- Menu-driven interfaces.

By directly using the commands, the command line is that interface that allows the user to interact with the computer. But there is an issue with it that the command can be understood by the computer only and also the commands cannot be changed, as they are fixed.

An interface that allows the user to interact with the system is known as the graphical user interface, as this is termed as the user-friendly and easy-to-use interface. For all types of users, this is attractive and also includes graphics and pictures. The command line is a black and white interface. As, this interface uses windows, icons, menus, pointers, so it is also known as WIMPS.

It has been seen that in the mobile phone or tablets, menu-driven interfaces are being used. To select the option of their own choice, this gives the menu to the user. For example, ATM is also the menu-driven interface. And not only the ATM, but many other devices use a menu-based interface. To select the options user has many options available on the screen.

It has been seen that mobile phones are small but they work like a computer, as they are being in different sizes and specifications. The interface of the system is not only being seen by the size of the mobile. Where the user and the computer can interact with one another is known as the user interface. In the showing content the screen size and determination limit in the cell phones.

For mobile phones, the exact values for the phone user interfaces do not exist, and some applications are also available for the desktop. In the physical mobile phone, these issues include fixed screen, lack of feedback and computers are different in from each other.

For both, these are different interfaces, but most of the demands are common in both of them like Wi-Fi, Bluetooth, photo gallery, engine, and many more. Computers are large and they are stationary, whereas mobile phones are small in size and they can be moved anywhere while users are walking, traveling, etc.

7.2. USER INTERFACE DESIGN BASICS

On anticipating user interface (UI) design focuses, what users might need to do and also making confirm that the interface has elements that are easy to access, understand, and use to facilitate those actions. From the interaction design, visual design, and information architecture, UI brings the together concepts.

7.2.1. Choosing Interface Elements

With the interface elements users have become familiar with which are acting in a certain way, so try to be consistent and predictable in your choices and their layout. With task completion, efficiency, and satisfaction doing so will help.

Interface elements include but are not limited to:

- **Input Controls:** buttons, text fields, checkboxes, radio buttons, dropdown lists, list boxes, toggles, date field.
- **Navigational Components:** breadcrumb, slider, search field, pagination, slider, tags, icons.
- **Informational Components:** tooltips, icons, progress bar, notifications, message boxes, modal windows.
- **Containers:** accordion.

For displaying the content, there are times when multiple elements might be appropriate. It has been seen that it is important to consider the tradeoffs when this happens. For example, sometimes elements that can help save you space, but more of a burden on the user mentally by forcing them to guess what is within the dropdown or what the element might be.

7.2.2. Best Practices for Designing an Interface

From knowing the users' everything stems, including understanding their goals, skills, preferences, and tendencies.

Everything stems from knowing your users, once the user is being known, make sure to consider the following when designing your interface:

- **Keep the interface simple:** To the user, the best interfaces are almost invisible. They avoid unnecessary elements and are clear in the language they use on labels and in messaging.
- **Create consistency and use common User Interface elements:** In the UI by using common elements, users feel more comfortable and can get things done more quickly. To create the pattern in language, it is also important to create layout and design throughout the site to help facilitate efficiency. Once a user learns how to do something, they should be able to transfer that skill to other parts of the site.
- **Be purposeful in page layout:** On the page, there is the consideration of the partial relationships between items and structure the page based on importance. To the most important pieces of information, there must be the careful placement of items that can help to draw the attention of the information and can aid scanning and readability.
- **Strategically use color and texture:** From the items using color, light, contrast, and texture a person can direct attention toward or redirect attention away.
- **Use typography to create hierarchy and clarity:** There should be careful consideration of how a person can use the typeface. Different sizes, fonts, and arrangement of the text help increase scanability, legibility, and readability.
- **Make sure that the system communicates what's happening:** Always inform the users of the location, actions, changes in state, or errors. The use of various UI elements to communicate status and, if necessary, the next steps can reduce frustration for your user.

- **Think about the defaults:** By carefully thinking about and anticipating the goals people bring to your site, an individual can create defaults that reduce the burden on the user. This becomes particularly important when it comes to forming design where you might have an opportunity to have some fields pre-chosen or filled out.

7.3. LITERATURE REVIEW

The software or application usability is the ease of understanding. Increasing the usability of the software or application user interface plays an important role. The medium of human-computer interaction is the user interface. For the user, the application is not fulfilling the requirements of the user, and then it is failed.

There is a different kind of user due to the less cost of the products. In the good cost, some people want a good product, but some of the people want the good product at normal or less cost. It all depends on the requirement of the user. There is an increase in the requirements or demands of the user due to an increase in the diversified users.

There are some of the challenges for designing those interfaces which have become more complex. These design issues are as following:

- User characteristics issue;
- User interface type and design issues;
- Message construction issues;
- Display graphics design issues;
- Look and feel issues;
- Performance issue;
- Language barrier issue;
- User experience issue;
- Complexity in UI design;
- Guideline issues;
- Navigating the interface;
- Organizing the display;
- Interaction design issues.

The person who is going to use the system a user is a person in the user characteristics issue, the user is not always the human and this can be the human, the machine, or any other living thing. It depends on the type and identity of the user.

For a developer, this is the main issue, if in the case the developer knows that for whom the developer is developing the product, and according to the user, the developer has to make it accordingly. This issue can be resolved only if the developer can observe the user first to think that what type of product should be good for the user. The user can be of different types:

- Nave user;
- Sophisticated user;
- Specialized user.

The system user who is having very little link is a nave user; for example, customer of the bank.

A user that uses the system and also comfortable in using that system is known as a sophisticated user; for example, manager and accountant of the bank.

Figure 7.2: Basics of user interface design.

Users who are experts in using the system, administrating, and maintaining the system are known as a specialized users; for example, administration, a system maintenance engineer.

There are two types of issues in the user interface type issue. The first one is the command line, and the other one is the graphical user interface. But today, the trend is being changed from the command-based user to the graphical user interface.

The user is being attracted because of the colors and the images in the graphical user interface. But if the developer gets the complete requirements about the interface that what type or style of interface is demanded by the customer, this issue can be resolved.

All the requirements should get clear to the customer as well. To solve such problems, the prototype is being developed because this is the demo of the complete system. The client then justifies that this is suitable for him or not, when this is being shown to the client.

Between the user and the system, the message is the source of communication in the message construction issues. In the user interface, this is the main issue. So, because of the construction of the message, this is the issue. The construction means the size of the message, font of the message, message type, error message, and the content of the message.

It has been said that the size of the message should be normal, it should not be very large or very small. According to the user, the message should be decided. From the windows error message is also the message. For the user, this is always useful, because if the user clicks on the wrong button or not following the correct instruction, then this is helpful to notify the user of damage. By having the message construction, these issues can be resolved according to the user error message notifies the user, when the user faces any issue.

The author emphasizes two terms in the display graphics design issues. One is the organized items and the second one is the use of color and the third one is a visual delight.

To the user look means that how the interface looks in the look and feel issue and in feel the response of the person when a user looks the interface. For the user, this easy-to-understand display and its working there is not scattered, in the case if the data is being organized and as a google design.

With the system, the user is using the application, but the application is not compatible with the performance issue. In some of the cases, there is an issue in using such an application because this is not suitable.

For the whole system, there is the fixed language that is being used in the language barrier issue. For the user, if the language is not understandable, then this creates the issue. To change the language to understand the application like that in google the solutions is that there should be the option.

All the problems are being listed above in the user experience. There are some of the problems, which include performance issues, interface issues, language barrier, and many more. So, these are the problems that are mostly being occurred.

It has been seen that the user interface design and coding can be complex in the user interface. To sketch the user interface, the application is being announced. To sketch the interface there some of the specific keywords. This is like the paint which is being available in the windows. Of the complete interface, the person can draw the complete sketch. So, to drag and drop the items allows the user and along with this to arrange the items of their own choice.

All the requirements should get explained by the user in the guideline issues and the exact guidelines should be followed by the software engineer. From the last time, to guide the effort of future designers' interface designers have tried to write down the guidelines.

To use the instruction in the interface, there is a fixed pattern in the navigation interface. So, there are some the fixed rules and principles for a user to install or uninstall the application. For this process, the user has to follow the following rules.

Navigating means to make the system with the fixed boundary and principles. Rules which are being of their nothing can be changed. In a proper sequence, everything happens. For example, short-cut keys for copy, paste, undo, etc.

With the special cases display design is a large topic in the organizing display. For the data display, there are high level with guidelines.

The user cannot interact with the system in the interaction design issues. The user doesn't interact with the system, as the user doesn't understand the system clearly because of the different issues like bad contrast of colors, scattered data, bad interface, message construction issue, etc. So, while making the software the designer should understand all issues.

7.4. CHALLENGES UX DESIGNERS FACE AND HOW TO OVERCOME THEM

People often think that user experience (UX) website design is optional when it is a must-have for companies that want to thrive.

Basically, by company culture, organization, workflow, or methodology, the biggest challenges that user experience designers face are being imposed. To the low access to data and poor internal communication, others are being related, which leads to a lack of vision around the products and services they are working on. Apart from this continuous learning, improvement, and good communications being required by the user experience roles.

So, after this, there are also some of the challenges that a UX designer will likely face during their career and ways to overcome each one.

7.4.1. The Ever-Changing Landscape of UX Job Titles

The ever-changing list of the user experience job titles is being one of the most common challenges which are being related. To name just a few, there are the user experience writers, user experience researchers, usability analysts, and information architects. It is really important to stay informed, to get a position in the industry.

To get a feel for the responsibilities of various job descriptions, a good idea is to read a web design hiring guide. Creating the personal brand will provide help, and it will increase the chances of getting hired faster.

7.4.2. Conducting Research with Limited Resources

To complete this important step, conducting user research is always recommended, but sometimes limited resources can inhibit a designer's ability

The good news is that to solve this problem there are several ways:

- There is a method that is considered as an inexpensive research method that is online surveys, through which questionnaires are sent to multiple participants. Nonetheless, building the questionnaire and analyzing the results require a serious amount of time.
- To assess the ideas and needs of the user, there is an informal discussion with the user. The demographic diversity should be there, and the key areas should be decided in advance which have to be discussed. The report from the Nielsen Norman Group will provide all the guidelines if an individual is wondering how to recruit participants for the focus group.
- When the team needs fast validation of several initial ideas by going to places, guerrilla user experience research is ideal where the audience will have time to listen and help, such as in a cafe or a park. By asking feedback from colleagues who aren't designers, engineers, or product people, this type of research can also be done at work.
- Parallel design is a method that requires the concurrent availability of the design team members. To build a new generation of ideas, they create alternative designs in parallel and incorporate the best elements of each design.

Figure 7.3: Conducting research with limited resources.

7.4.3. Organizational Responsibility for the User Experience

As a separate entity user experience, designers shouldn't view themselves. It can be difficult because of the organizational boundaries, but with the web designers and developer user experience, designers should endeavor to work closely, to mutually influenced toward a unified mindset to create optimal results for users.

7.4.4. Artificial Intelligence to Change the Role of UX Designers

By the artificial intelligence (AI), the face of many industries has been changed, but the question also is about AI's impact on the design industry. Along with these designers are also afraid that they will lose their talent to the advanced technology.

In general, there is a study by Pfeiffer which shows that creative professionals are confident that they will be able to integrate the AI-based software tools into their creativity, while 62% of design professionals believe artificial intelligence and machine learning will be very important to their creative jobs.

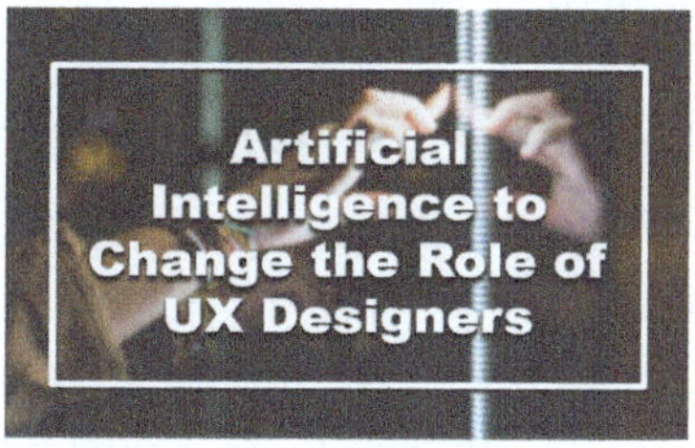

Figure 7.4: Artificial intelligence to change the role of UX designers.

The title may get the change to the behavior designers, who will provide data to AI algorithms through a set of parameters by the user experience designers. For example, AI can be used to run A/B tests automatically, analyze the results, update the design accordingly and restart the process.

7.4.5. The Broad Role of a UX Designer

The full lifecycle of the customer experience is being influenced by the user experience, so it means more than a nice design and pixels, adding pretty colors, and drawing some buttons. To improve the relationship between a company and its customers, the goal of user experience design is there and along with this to improve their experience with the services and products that are provided.

Therefore, it has been seen that the set of skills of a user experience designer is complex, so to study the user, an individual needs to understand the business and its position in the market and also the goals to create better experiences.

For solving the problems, a practical way is to overcome this challenge is to always come back to the tried and tested 4-stage UX process:

- Discover;
- Define and validate;
- Define and test;
- Develop and measure.

But it has to be kept in mind that this doesn't mean there is an exact recipe or perfect user experience, just as Don Norman said. As a part of the large system what an individual is doing. All the factors have been taken into account and also the people that are part of the system.

With the use of the system, an attempt should be made to understand how it works and trying to get it to do what is being cared about. While working on the user experience design, there is always a thought that what person is trying to accomplish, how they will appreciate it, be happy with it.

7.4.6. Career Shift from Web Designer to UX Designer

There is a fact that the user experience design has a huge impact on a business, apart from the financial reward of making the switch into user experience design. In the US, web designer earn about $46,000 per year, while user experience designers earn $74,000 per year, according to PayScale.

Essentially, when moving to a user experience design, there is a great benefit of having a web design. As collaborating with colleagues will be easier if that person who wants to become a user experience designer has a web design background. Along with this, intense collaboration, communication plays a crucial role which is being involved in the user experience design, so speaking the same language will help.

Figure 7.5: Career shift from web designer to UX designer.

To become a user experience designer is a good idea or not. It is important first to know what web design and user experience design have in common: building interactive prototypes or visual designs, as well as doing user research and testing. They're some of the other aspects that the two disciplines have in common are creating an emotional design, solving problems, and multidisciplinary.

7.4.7. Enterprise UX Challenges

An individual has faced much more complex situations as working as an Enterprise User Experience Designer, and they also have some different needs. As design has to deliver scalable features, there will be more responsibility and pressure, and it should increase product development efficiency and efficacy by at least 25%.

There is another challenge that dealing with the misconception that user experience means only to improve the look and feel of a product when it is more than that. Or, the end stakeholders believe they already know the users' needs.

Moreover, among designers, developers, testers, and product managers there is poor collaboration which can lead to a lack of communication and finally to the bad design and internal conflicts. To avoid complications, a good thing to do is to integrate a focus on design thinking in the company, while it's off to see things differently, which will lead to a more collaborative environment and employees feeling more empowered. Plus, using the principles of design thinking will result in better design of products or services.

7.4.8. Consumer Psychology and UX Design

In today's market, consumers have so many options, which makes it difficult for user experience designers to know what they want. User experience designers need to employ consumer psychology, by focusing on user satisfaction and making the decision process easier.

There are various types of behaviors that can get triggered by the colors, spatial structures, as well as contours and shapes, motivations, and emotions and which can influence consumer behavior subliminally. Consumer psychology is being understood by a user experience designer and who also knows the business side of the product or service will be more successful.

But another question is that what can a designer use from psychology and bring to user experience. In understanding the user more profoundly, there are several theories about emotions, motivation, perception, and biases that are a great advantage.

UX designer plans to attract based on the type of the consumer, the user will incorporate specific colors and structures. Such as neutral, structured, and curved designs although users have general preferences, to increased customer satisfaction or profit, there is still little proof that user experience design and psychology are directly related.

Figure 7.6: Consumer psychology and UX design.

However, there are multiple psychology principles that all user experience designers should know about:

- To perceive them as a whole, subconsciously gestalt principles of visual perception states that users group separated objects.
- Before taking action on the website Psychology of Persuasion explains that viewers have to be convinced.
- Hick's Law mentions that users want to find what they need on a website extremely quickly.
- Psychology of Colors is the science of how human behavior is influenced by color.

- Von Restuff Effect is also known as the "Isolation Effect" and states that distinctive items are more easily remembered than ordinary ones.
- Psychology of Shapes explains that we form subconscious associations between shapes and certain qualities.

7.4.9. Working in Scrum Teams

To integrate the user experience design, there are various approaches within the Agile methodology. With the development sprint running a design sprint happening in parallel and along with a development sprint and a mixt there is a separate design sprint which includes the developers and the user experience designers and testers.

A study is being conducted by the nongroups which shows that Agile user experience works perfectly if management understands and takes into consideration the value of user experience. A designer will face several challenges as a part of the integrated scrum team like organizational, communicational, and the lack of time.

There is one of the most important recommendations for user experience designers and developers under these circumstances to have and maintain a healthy relationship by sharing responsibility, ownership, and learning from each other. For example, user experience designers should require web developers to not implement anything without first doing thorough user research, and at the same time, developers should not give user experience designers direction of their work.

7.4.10. Receiving Negative Feedback

From the design managers CMOs, CEOs, product manager's approval is required for a UX designer and sometimes even from the designer peers. UX designers receive feedback is the most common challenge and that sometimes is rather subjective even if it comes from design professionals.

Getting feedback that is based on intuition can be frustrating, but to the design, no changes can be made. There is a thought that at least, it may sound like a ridiculous question, but it helps an individual to get rid of vague opinions and start a professional discussion.

Therefore, for the design choices and engage in conversations, a person should also need to make a strong case. When your design is being discussed, which is to be kept in mind, and there some of the colleagues who express their opinion and create a thriving service or product, they all try to achieve the same goal.

7.4.11. Design for the Future

There is UX design that is undergoing a revolution right now to predict the future though it's quite difficult, and there are some of the directions in which it will be heading. For accessibility one of them is designing.

About this gap, companies have been more aware lately, and as a result, some products are already there, and some services help people with disabilities that will have a better user experience. For the UX designers, the challenge will be to put themselves in the shoes of those users with disabilities, interview them and discover what their main needs and issues are.

From the recent years, there is a consideration in the rise of mobile and wearable devices, as well as virtual reality, augmented reality and artificial intelligence, another direction in which UX design could also be heading is design-agnostic user interfaces.

There is an effort to adapt this future variety, as UX designers will create with a multitude of devices in mind, and to test carefully it will be crucial and to obtain accurate and effective answers the right question needs to be asked.

Within the company to overcome all these challenges, the role is being defined for right now or whether an individual making the decisions about the future career, and it's always being important to research and stay informed. A UX designer and individual must be aware of the most common challenges and will be able to identify them much more easily and take appropriate action right away.

7.5. THE BIGGEST CHALLENGES IN DESIGN

Over the saturated marketplace, a logo design is being created that stands up in an increasingly, or the art of working from home is being mastering, and from the last 20 years, it has been seen that there is a lot of change in the design world which brings a new set of challenges.

Figure 7.7: The biggest challenges in design.

7.5.1. Working within Fast Turnarounds

It has been said by the Kirsty Carter of APFEL that communication is constantly speeding up and people are becoming more and more connected with one another. So, the consequence for this is that the immediate answer is being expected and likewise the design solutions. To work within the accelerated turnarounds new ways, need to find. To save time on certain things is being allowed by creativity and innovation, and to slow down this will help to buy more time, for example, for research, reflection, and investigation. It has been seen that the friction is being created by the challenges, and this will often provoke creativity, so from the design and creative industries, the biggest challenge could even better work.

7.5.2. Balancing Thinking with Doing

Heather Stern of Lippincott points out that the pervasiveness of design thinking at the C-suite level is as exciting as it is challenging. It's wonderful that the world now understands design in a bigger context, but people mustn't lose the unique facets of design beneath it. The challenge is balancing the emphasis now placed on the thinking with the actual doing.

7.5.3. Staying Relevant

It has been said by Studio Output's Dan Moore that the near future is about defining. So, everyone needs to be at the forefront of digital and understand how marketers are currently using the technology, so a user or a person can work it out, if he or she wants to fit in this new world.

7.5.4. Adapting to Technological Change

Anticipation is there regarding the technological change which is being adapted, master it, excel, and in a very short time then be open to new influences.

7.5.5. Making Great Work

Over the last couple of decades, it has been seen that the main challenges haven't changed too much, as is suggested by James Wignall, so according to this, it has been assumed that the course of the next 20 will remain pretty similar. The main challenge is always to make the work great, stay relevant and stand out even the technology evolve and software changes.

7.5.6. Staying True to Design Fundamentals

Designers must remain true to the core conceptual ideas in an ever-evolving commercial and social landscape while keeping up with the technology and customer engagement. To try different approaches, they must be brave enough, flexible enough to adapt swiftly, but always appreciate the importance of doing the job well. As it has been seen in many start-ups that keeping these values as their core philosophy is the way to success and which shows that there always will be a need for these fundamentals.

7.5.7. Having a Holistic Perspective

Nowadays it has been seen that an isolated view is often the result and the creative industry favors specialization. There is a belief that with a holistic perspective, there is room for a generalist. To make the true advancements is being expected from them.

7.5.8. Being Unique

Styles and trends can be replicated in an instance if everyone can access the same inspiration as everyone else. Brand values, integrity, honesty, and originality will become more important than ever.

7.5.9. Being Multi-Skilled

From the school of visual arts, Richard Wilde points out that students entering the creative industry not only need expertise in design and typography, conceptual thinking, and execution skills, and along with this today they also need expertise in a variety of disciplines, including motion graphics and interaction design, which is being supported by the latest technology, and coupled with strong presentational skills.

7.5.10. Cultivating Human Experience

The line will be blurred between technology, human experience, and creativity. With the advancement of technology, certain skills will no longer be needed as computers and automation make aspects of 'designing' accessible to all. By finding new ways to embed experiences, memories, stories, and culture into creative output, the creative industries need to cultivate the human.

7.5.11. Encouraging New Talent to Aspire to Greatness

For the graphic artist to flourish the marketplace no longer provides a substantial enough platform. The real challenge is to find a non-web-based portal for its exhibition, as there is still an overwhelming amount of talent out there. To allow for the showcasing of the best talent, dwindling budgets need to be realigned and also to encourage new talent to aspire to greatness.

7.5.12. Staying Hungry for Other Design Disciplines

For the other design disciplines, creative need to ensure that they stay hungry. They need to cultivate a sense of wonder about many things, even if their final product is purely digital, from sculptural, natural, and man-made forms to product design, interiors and exteriors, audios capes, and lighting effects. In a connected world, design experiences of the future will touch many of these, but a digital product shouldn't mean a shortcut to a shallow solution.

7.5.13. Maintaining Personal Interactions

To discuss a brief or a creative issue, one of the downsides of the internet revolution is that it has nurtured a generation that doesn't get on the phone or meet a client in person. As, the entire rely is on email and social media to communicate, which can be very one-dimensional and open to interpretation. Personal communication is the foundation of creativity and good business, and an idea can often be discussed better verbally, which inevitably leads to better solutions.

7.5.14. Finding the Right Clients

The problem of finding a good client is the biggest challenge which is faced by the designers and good clients mean actually who value good design for themselves and clients who don't perceive design as just another expense, but a worthwhile investment.

7.6. USER INTERFACE CHALLENGES

Creating user interfaces (UIs) was a simple task just over a decade ago. To create a simple MS-DOS interface developers used whatever their tool of choice provided. Since then, things have gotten considerably more difficult. Developers face a plethora of choices that are supposed to help them to fulfill those expectations, and the user expectations have risen sky-high.

Developers spend a lot of time training at EPS, and there is a training program that is run regularly is called Ask Markus. The idea is that developers can (anonymously if they want) submit questions about things they encounter in their daily developer life (and yes, the question has to be development-related).

Figure 7.8: Various challenges have been faced in designing the user interface.

7.6.1. The Available Options

For the UI layers, there are surprisingly many options these days. Based on the Windows Forms technology, there is a rich Windows client. Today everyone not only uses what is available out of the box, but technologies like GDI+ have become so accessible that anyone just can't help but use them in advanced UIs. There is another variation on that theme which includes Windows console-based applications, although it is not sure that whether this advanced interface technology is being considered or not.

Then after this, there are some of the HTML-based web applications, and although ASP.NET, which is yet another UI technology and it also shields the user from much of the low-level HTML. Without knowing the HTML, it is not possible that any half-decent web developer could get away and along with this related client-side technology like JavaScript.

As there are some new techniques available like AJAX, that try to overcome some of the shortcomings of Web browser clients, so this the only half the Web-development story. And the user or an individual will probably get into situations where the user will think seriously about how the content is generated on the server, only in the case if the user is building advanced Web applications.

The end-user will end up creating the custom server controls, which is not so difficult as it sounds. On the other hand, there is a need to display content that matches the definition of a document more closely than that of a control. Users will use the style sheets to make Web applications look good.

Then there is the traditional issue of design, which means battleship gray Windows UIs are the domain of the developer. It has been seen that with the more recent generation of Windows applications, things are not quite clear, and it also has a heavier artistic influence. In Web applications, this trend is even more obvious, where the use of images and design is much more common.

For the code-behind model, that was the reason and where-at least theoretically developers could hand over the ASPX page to an artist and get the design done and ultimately incorporated in the Web application. This didn't happen very often. In Web applications, the need for artists and designers is very real.

For the artists and designers, it is pretty obvious that another technology makes the need for them, such as Macromedia flash. These days it has been seen that almost all browsers can display the flash content statistically, and there are a lot of web pages that are taking advantage of this.

From a developer's point of flash is worth checking out, and it may be lacking a bit. So, another question is there for the developers what makes flash a bit unusual and about the concept of animation because most developers do not (yet) think about their interfaces in terms of timelines.

But further, it has been seen that relatively in the near future, there is another paradigm shift in UI development that is being faced by the user. There is an Avalon UI technology that will ship with WinFS and Windows longhorn. It is much closer to DirectX (Direct3D) than the conventional GDI-based interfaces.

There are some of the key features of Avalon, such as animation and the ability to create objects in 3D spaces. There is a new syntax which is known as XAML and also used by Avalon to declaratively define interfaces. To define the interfaces XAML is like a specialized XML. The overall approach is not entirely unlike ASP.NET pages today. Except, of course, that XAML and Avalon provide rich-client Windows features.

There is a need to have the designers and artists continue with the Avalon. To the Avalon experience, the concept of a polished user interface is central. This means the artist who understands concepts like animation, timelines, and storyboards are needed, and not only have those who understand only fancied graphics.

7.6.2. What to Focus On

After analyzing all this, still, a question left without the answer that on which technology should a developer focus.

All these technologies need to be understood by the developers in any way. They're some of the technologies which are get ignored by the user like Pocket PCs, Windows CE, Tablet PCs, Aux Displays, Smart Phones, SPOT, and others. But for most developers knowing all the technologies mentioned here is not a realistic short-term goal. So, there are some of the suggestions which are available for the developers which they might use as a roadmap.

With windows, web developers should familiarize themselves and vice versa. The role of windows and web developers is not being separated. There are very few applications that are Web applications or pure Windows applications. By the developer, the focus is on one of the two, but the other is usually present as well.

To tackle the Windows side as well user or the developer must have enough knowledge if a user already knows enough .NET to build an ASP .NET solution. Learning HTML and ASP.NET shouldn't be a huge hurdle either if a user is a Windows developer and has been a Windows developer for a while.

XML and the related technologies must be known by all the developers. Because too much XML is involved in development these days to ignore it, besides it is not all that hard. Along with this, there are some other related technologies which also need to be understood, such as XPath queries and XSLT transformation.

Without understanding these not maximum benefits can be achieved by XML and get only the fraction benefit of XML. In all the aspects of UI development knowing these technologies will be useful. It is surprising all the time by knowing the number of developers who are not familiar with XML. With declarative programming, all should get familiarize. Some people love declarative programming and others just hate it. As there is so much coming almost in all areas of development and particular in UI development?

Learning HTML is done by the individual years ago. As, HTML is ASP. NET development today, and HTML-based applications are not going away. In addition, understanding the HTML approach will be helpful with a lot of new technologies. Sure, XAML is not HTML, but if a user or the individual understands HTML, XAML will seem quite natural.

7.6.3. Check Out What You Can Do With Styles

There are many style sheet technologies. HTML's Cascading Style Sheets (CSS) are a good place to start. Sure, they are very primitive compared to what a user can do with styles in Avalon, but users have to start somewhere, right. So, moving on to XAML styles will be a lot of fun.

7.6.4. Start Looking into Animation

Control properties can get changed by the user in windows forms and even ASP.NET. From one state to another, an induvial or the user animate a property in Avalon. So, the animation will be important and it will be easier to learn it over time than all at once when the Avalon wave hits the user. A good way to familiarize yourself with the concepts of animations and timelines is through Macromedia Flash.

7.7. COMMON PROBLEMS WITH THE UX PROCESS, AND THE SOLUTIONS

About the variety of challenges, there is discussion done in this chapter which is being faced and how eventually overcame them. To enhance the user experience design process in smaller projects, many of these learnings can be applied.

Figure 7.9: Common problems with the UX process.

7.7.1. Facing Up to the Feasibility

To communicate the ideas to the client with a small project, which is going right into the wire framing tool of choice and making a quick interactive prototype and it is perfectly acceptable. Before getting to that stage, there is a more complex scenario where many people will map out user journeys, and this is nothing new. The feasibility of the project is uncertain when the user journeys are so complicated and their number so high.

Within the system, there are many users, and each user has many different roles, as in a web or mobile application, users might have different workflows that to carter the user. In some parts, complexity may exist, and it has been known that large chunks are likely to be feasible and the logging in, the registration form, and the navigation elements are all so familiar.

Logical steps of following the user's journey have been taken while wireframing, starting with the landing pages and then through to the ultimate goal. However, with so much ground to cover, cooking the side dishes simply doesn't make sense if the main course is unachievable.

On the market condition and the technical constraints feasibility studies in the business often focus, but determining whether a sound user experience (UX) can be delivered is just as important. If it is discovered down the line that the project is too ambitious, then a lot of effort might have been wasted.

7.7.1.1. Solution: Tackle the Challenging Parts of the UI First

It has been seen that during the discovery phase, the most challenging aspect of the user interface is to identify that and then create wireframes for those particular parts. So, this fits perfectly with familiar UX practices, such as mapping out the different types of users and their journeys through the system.

Addressing these challenges will help you to estimate how much development time is needed and will not only help the design progress faster.

By considering a hypothetical scenario, it can get explained. An individual is being asked to design a website containing a feature that enables a user to invite people to design a friend's birthday card, together in real-time.

7.7.2. Presenting an Incomplete Journey

This problem is considered as the side effect of the first solution and it has been left behind that how an individual will account for the gaps. It is required to share it with the stakeholders and potential users to validate the work done so far.

7.8. CONCLUSION

In the conclusion, this chapter explains discussed the various challenges and opportunities that have been addressed in user interface design. This chapter also discussed the basics of user interface design, such as choosing interface elements, and best practices for designing an interface.

In this chapter, several numbers of challenges that user interface designers face and how to over those challenges include the ever-changing landscape of user interface job titles, researching with limited resources, organizational responsibility for the user experience, artificial intelligence to change the role of user interface designers, the broad role of a user interface designer, career shift from web designer to a user interface designer, enterprise user interface challenges, consumer psychology, and user interface design, working in scrum teams, receiving negative feedback, and design for the future has been discussed.

Towards the end of the chapter, several numbers of various biggest challenges that have been faced in the process of designing such as working within fast turnarounds, balancing thinking with doing, staying relevant, adapting to technological change, making great work, staying true to design fundamentals, having a holistic perspective, being unique, being multi-skilled, cultivating human experience, encouraging new talent to aspire to greatness, staying hungry for other design disciplines, maintaining personal interactions, and finding the right clients has been discussed as well. This chapter mentions various common issues that have been coming across in the UX process, along with the solutions to those issues such as facing up to the feasibility and presenting an incomplete journey.

REFERENCES

1. Cass, J., (2020). *How to Turn UX Challenges into Success*. [online] JUST™ Creative. Available at: https://justcreative.com/ux-challenges (accessed on 6 July 2021).
2. Chris, (2014). *6 Common Problems with The UX Process, And 6 Solutions!—Smashing Magazine*. [online] Smashing Magazine. Available at: https://www.smashingmagazine.com/2014/05/6-common-problems-ux-process-6-solutions/ (accessed on 6 July 2021).
3. Julia, S., (2016). *14 Biggest Challenges in Design* | Creative Bloq. [online] Creativebloq.com. Available at: https://www.creativebloq.com/graphic-design/14-biggest-challenges-design-today-31619475 (accessed on 6 July 2021).
4. Markus, E., (2019). *User Interface Challenges*. [online] Codemag.com. Available at: https://www.codemag.com/Article/050083/User-Interface-Challenges (accessed on 6 July 2021).
5. Medium, (2016). *Five challenges the Interaction & User Experience Design Industry face*. [online] Available at: https://medium.com/@furthermore_ux/five-challenges-the-interaction-user-experience-design-industry-face-86e851ddd5ee (accessed on 6 July 2021).
6. Usability.gov. (2021). *User Interface Design Basics* | Usability.gov. [online] Available at: https://www.usability.gov/what-and-why/user-interface-design.html (accessed on 6 July 2021).
7. uxplanet.org. (2019). *5 Common Challenges UX Designers Face—And how to Overcome Them*. [online] Medium. Available at: https://uxplanet.org/5-common-challenges-ux-designers-face-and-how-to-overcome-them-c704f04eb952 (accessed on 6 July 2021).
8. Waleed, I., Muhammad, S. A. M., Shanza, T., Maha, A., Jawad, A., & Saad, S. M., (2018). User interface design issues in HCI. [e-Book] Faisalabad, Pakistan. *IJCSNS International Journal of Computer Science and Network Security* (8th edn. pp. 153–157). Available at: http://paper.ijcsns.org/07_book/201808/20180824.pdf (accessed on 6 July 2021).

CHAPTER 8

FUTURE OF USER INTERFACE DESIGN

CONTENTS

This chapter sheds light on the future of user interface design. This chapter also explains the current state of the user interface design. This chapter has tried to explain the development or rise of the voice user interface and its accessibility. This chapter provides highlights on how artificial intelligence will disrupt the user interface design. This chapter also explains various types of future user interface trends, along with the impacts of user interface design on everyday life. This chapter also addresses the concept of biometric and cybernetic interfaces.

8.1. INTRODUCTION

For decades now, User interface (UI) design has played a crucial role in computing. The concept of UI design first burst onto the scene when graphical user interface operating systems (OS) like Macintosh and Microsoft set foot in the marketplace.

In its nascent stage, digital icons corresponding to a particular file or program opening at a double-click by a user were the form of UIs. In that era, skeuomorphic design transformed and facilitated technology for mass use by designing icons mimicking their real-world counterparts. A classic example of this is the recycle bin icon used for discarding files.

Skeuomorphism makes interface objects familiar to the end-user by using images and concepts they recognize.

Today, we live in an age of design and continuously evolving user interfaces; a dominantly technological era where everything in our lives, be it clothes, accessories, homes, cars, restaurants, roads, and so on have become a huge user interface reaching us through our smartphones and other such gadgets enabling an uninterrupted and perfect interplay with technology.

UI design is the process to form interfaces in software or computerized devices that are user-friendly. With the latest advancements in UI, it can take the form of desktops, display screens, and keyboards. Additionally, it also reflects how the end-user interacts with an application or website.

Mobile and web applications have witnessed colossal growth in recent times, and smart mobile devices have seen a burgeoning demand in the market. This has further augmented the capacity and purpose of UI, enabling it to further enhance the overall user experience (UX). UI design also allows for competition amongst brands as it lends a distinct identity to each.

Figure 8.1: Future of user interface design.

Even in its preliminary stages, the development of UI design as language-independent icons and images, assimilated and organized various concepts like menus, windows, and pointers that have made navigation through any OS much easier. Our smartphones and tablets use the same idea for navigation and browsing.

Over the past few years, a more formidable connection has formed between UI and UX, and both exercise a powerful impact on each other. For example, product development teams that build highly responsive mobile apps and websites have to take into account both UI and UX equally as both have a major impact on the design of the end product.

The aforesaid scenario is responsible for the transformation of the designer from being the artist of the product to the strategist behind its conception.

8.2. THE CURRENT STATE OF UI DESIGN

Today, the latest fad in the industry is flat design; a minimalist design style laying emphasis heavily on the usage of flat colors and clean lines, making it appear more streamlined, a far cry from Apple's pervasive skeuomorphic style. Flat design is often regarded as a reactionary movement against the three-dimensional skeuomorphic style which was rampant before flat design.

This trend has influenced UI design by the application of flat surfaces, cleaner edges, and downplaying graphics, and toning down the design to a simple form. The evolution and popularity of the flat design trend exhibit and confirm a gradual shift within the industry to create designs keeping in mind the different form factors.

In sharp contrast to this, websites have amalgamated polygonal shapes, basic geometric layers, and bold lines with a clear intent of grabbing the attention of the audience and keeping them hooked.

In the recent past, tactile designs have attained rapid popularity. The unique characteristic of hyper-realistic appearance is the highlight of this particular trend. Apart from these trends discussed here, numerous examples can be seen of websites without any borders, without any multiple layering, with deliberate animation, and the use of large images.

Figure 8.2: The current state of UI design.

Soon, it can be inevitably expected that higher and highly competitive standards would be maintained within the app and web world. This would enable improved user interactions as a result of a smooth and consistent exchange of work between UI and UX.

8.3. LARGE IMAGES AND TYPE DESIGNS ARE HERE TO STAY

It can be safely assumed that for a brief time, those websites and applications that characteristically feature large images and type design will meet with increasing success and popularity.

The constant demand for engaging and capturing the attention of the audiences by the use of imagery that, very explicitly and arouses emotion while partially concealing necessary information explains the massive popularity of this design approach, particularly in the case of large images. Buzz Feed's mobile site is an excellent example of this approach.

Simultaneously, numerous schematic typography-based designs often contribute to being the predominant feature in UI design. By and large, these typography-based designs are the most favored choice among content-based websites and apps that fundamentally attempt to emphasize and draw attention towards the content put on display rather than the design.

Nonetheless, neither of the above-mentioned approaches is as easy as just attaching a high-resolution image or large texts. The decision of which approach to adopt is determined by striking a fine balance between research data and design aesthetics.

The near future may behold extensive usage of large imagery and texts, but it does not necessarily mean that these design elements will become static. With constant upgrades and revamping of video codecs, the coming years shall witness the ousting of static imagery in backgrounds and headers by videos.

8.4. THE RISE OF VOICE USER INTERFACE (VUI) AND ACCESSIBILITY

Siri and Alexa have laid the foundation for the imminent future and what it holds. In the future, almost all mobile apps and websites will be VUI enabled to heighten audience engagement.

Routinely, it's much more effortless to ask a question than pick up the device and click through it until one is lead to the right answer. Consequently, tremendous growth is expected in screen-less user experiences, triggering a major leap in the incorporation of apps being more fused into our daily lives and becoming an essential part of our routine activities, be cooking, cleaning, or driving. In today's world, VUIs have earned a well-deserving reputation of being ubiquitous and all-pervasive, thanks to the meteoric rise of the Internet of Things (IoT).

Figure 8.3: The rise of voice user interface and accessibility.

According to breakthrough research carried out by Gartner, the world's leading research and advisory company, virtual personal assistants (like Alexa and Siri) will dictate as much as 20% of user interactions by 2019. Furthermore, it is predicted that the following year can also expect 30% of online sessions not to feature a screen as virtual assistants communicate all the necessary information.

Although, it sounds peculiar to discuss design in the absence of a screen, yet if carefully approached, the thought process of UIs in terms of functionality and UX is more or less the same. Regardless of this feature, it does not necessarily mean that graphical interfaces are going to die out soon.

8.5. ARTIFICIAL INTELLIGENCE (AI) WILL DISRUPT UI DESIGN

The remarkable advent of AI in recent years has led to a lot of speculation and rumors across industries about how it was going to replace humans. News stories mostly focused on gauging its impact on factory jobs and back-office functions like data entry, but it seems that AI is also about to unsettle the design world.

Notable changes are going to be made as AI is going to revamp UI and UX for both Interaction Design and Visual Design. It is anticipated that soon, many interfaces will not be designed by humans. Instead, the designing job would be done by software.

The two primary ideas navigating this change are Perfection of Interaction Design and Personality Responsive Design.

Figure 8.4: Artificial intelligence will disrupt UI design.

8.5.1. The Perfection of Interaction Design

The perfection of interaction design focuses on not "reinventing the wheel" with every new design iteration. For example, there's no real need to change the date picker with every update.

Websites and apps have been existing for a considerable number of years now, and our growing body of knowledge around user testing has witnessed marked growth. The vast professional domain now allows for narrowing it down to what's best for a variety of use cases. Thus, unless a remarkable alteration or shift takes place in the user's content or the nature of the interface, there won't be any need to change it.

Over time as AI becomes more intelligent, the rapid generation of application interfaces takes place enhancing user interaction, but these won't get designed by humans. Interaction design is all about the interactive components and graphics stored in UI/UX libraries within the system and leveraged by AI whenever it can optimize individual experiences.

The perfection of interaction design suggests that the AI approach to UI design should concentrate on the interaction paradigm in real-time and strive not to deliver great designs, but the right design, most of the time.

8.5.2. Personality Responsive Design

Personalization or customization is the key element in all modern design and digital strategies are all, and AI is expected to take that to the next level with fluid AIs. For example, based on user behavior, the UI layers of any digital product can re-render in real-time to accommodate the end user's unique preferences.

The process of drafting, organizing, and developing apps, software, and even websites will undergo a massive transformation as personality responsive design comes into the picture, as the focal point will shift from what the end consumer might need or want to what kind of a relationship a particular brand wants to maintain with its target audience.

8.6. DATA WILL DRIVE THE FUTURE OF UI/UX DESIGN

The introduction of AI within the UI/UX design space has changed the dynamics of the situation with AI now guiding UI/UX design. This has changed the impression about design as the earlier design was distinguished and identified as an art form. However, with the advent of UI/UX in design, it has converted into something more scientific like Google's material design.

The whole dynamics have undergone a massive transformation, with designers operating as data scientists undertaking the responsibility of carrying out controlled experiments to collect data to decide the finest technique to present information. Leveraging data is also being carried out, enabling companies to identify the best way to help users navigate through a variety of interactions within applications.

In today's world, data analysts and digital strategists analyze and evaluate websites and apps to estimate success and recognize new and promising opportunities. In the coming years, this particular aspect of UI design will also get automated, and all the research and implementation would be undertaken by AI.

While it's certain that in the coming year's data and AI are likely to play a dominant role in UI/UX design, it is a matter of concern as to what would the future hold for designers? While it's difficult to make any reliable predictions, we can make an educated guess.

A careful study of the evolution of UI design informs us that it is an ever-evolving that undergoes changes continually thereby rendering designers to be highly adaptable to cope up with the demands of the business and the end-user. Thus, soon, one can expect the same to come and also a possibility of the emergence of designers who are also coders.

With AI playing a key role in UI/UX design in the future, the role of designers will become highly fragmented and specialized, although the titles like "UI designer" and "UX designer" will not be dying out change any time soon. Data in design further enables the designer to become more flexible and multifaceted to research, test, code, and adapt wireframes and MVPs.

Stiff competition in the market demands designers to stay abreast with the latest developments and pay more attention to detail to differentiate digital products and stay a step ahead of the competition. With data playing a critical role in design, it also determines how the UI adapts to user behavior, how well it can anticipate the user's needs, and how fast it can get delivered.

Designers with a strong eye for detail within this capacity should undertake procedures that bring about product enhancement and continually deliver them at a micro-level based on app data and overall market trends. The same is true when it comes to AR, VR, and VUI interfaces.

AI in future will zero down its focus completely tactical design projects, making room for someone to design the AI systems and their personality models. This could turn out to be the opportunity that would lead designers down the programming path by catering to these personality types and designing after them.

The issue of security must be at the helm of the minds of UI/UX designers while they work with interfaces. The advent of all these IoT and AI-powered technologies and their ever-growing role in our daily lives makes it even more critical for UI/UX designers to be more cautious in terms of security and privacy.

8.7. THE TRENDS AND PROPHECIES OF UI DESIGN IN FUTURE

Simplicity, flat design, and minimalism have garnered a lot of popularity in 2018 and 2019. What remains to be seen is whether these trends continue to be as popular as they were earlier in the coming years? What also is highly anticipated is whether minimalism and flat design pass the test of time or will they too go the way of skeuomorphism circa early 2008? Or, with nowhere else to go, will realistically design metaphors once again take center stage. 2019 saw the meteoric rise of certain companies and products as also the complete downfall. Some factors influence the success or failure of a product, the most important being user experience. 2019 has taught all organizations and the design world that the design and user experience of any application must be the topmost priority.

Figure 8.5: The trends and prophecies of UI design in future.

8.8. SKEUOMORPHISM

In simple terms, Skeuomorphism is when digital interfaces seek to mimic real-world interfaces. A product that popped up first in the 1980s, with the emergence of easily recognizable and realistic digital metaphors, for instance, the trash can, and save icons.

Skeuomorphism dominated the scene up until 2008. This year, Apple released a new version of iOS, which was focused on flat design. Within a short time, the entire design world followed suit and shunned skeuomorphism and switched over to flat design. Flat design was an extremist reaction that came up because of the heavy overuse of Skeuomorphism in design. Yet both Skeuomorphism and flat design face heavy criticism. Skeuomorphism is great as it evokes familiarity in a user. They are already aware of the functionality of a trash can, so the jump to a digital trash can is easy. But in certain instances, when interfaces start to become bogged down by realistic leather textures and ripped paper, it's starting to be a bit much.

In contrast, flat design does a great job of simplifying interfaces, but sometimes they turn out to look too basic. The absence of shadows or realistic metaphors can make it troublesome for a user to comprehend without basic training. Both the styles shall remain in use, but their usage needs to be more user-friendly and appropriate. For example, smartwatch interfaces benefit from being Skeuomorphic while mobile phone stock apps benefit from flat, clean design.

8.9. THE PAST, PRESENT, AND FUTURE OF USER INTERFACES

User interfaces are the face of a computer system, i.e., it is the space where interactions between the machine and humans occur. It has undergone tremendous changes over the past few decades. Therefore, it can be established that UI plays a crucial part in deciding the future of computers. The very concept and theme of UI came into being when Graphic User Interface (GUI) operating systems such as Microsoft and Macintosh ventured into the market.

Major advancements have been witnessed in the last few years in terms of digital interfaces. Though they have gained ground, there's still a lot more that needs to be achieved as there are many potential directions of growth for future interface designs. We're already seeing many new trends crop up, and it's exciting to think about how they'll change our lives.

UX has gained prominence ever since it emerged on the scene and the potential impact of UI and UX on the process of design in product development is another major phenomenon in the last few decades.

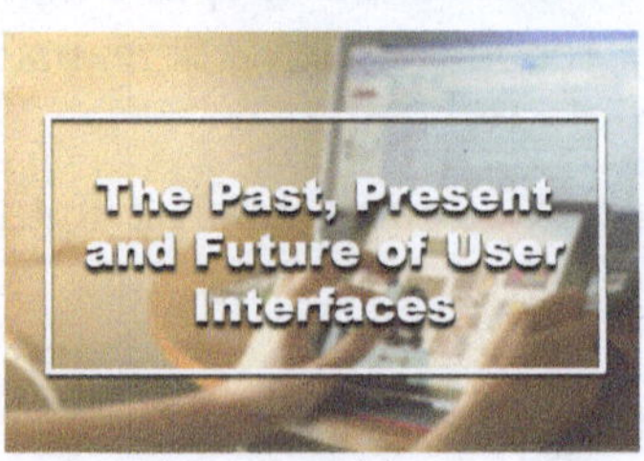

Figure 8.6: The past, present, and future of user interfaces.

Under this topic, the history, current scenario, and potential capabilities of future interfaces would be discussed at length in the chapter.

8.10. HOW DID INTERFACES EVOLVE?

The world's first Personal Computer (PC) was launched by IBM in the year 1981, and it came along with Microsoft's MS-DOS operating system that required users to key in command prompts that would execute specific tasks on their PC. It was dull in appearance, monotonous, and required users to memorize a lot of commands to complete tasks.

With the launch of Apple's Macintosh, in 1984, the entire environment of computers turned around. The Macintosh was a computer that popularized the use of Graphical User Interfaces (GUI) on personal computers. It took everyone by storm, as it popularized the usage of computers. Users were not merely using their PCs just because they had to out of some obligation, but because they wanted to, as it revolutionized the interface.

In the coming decade, displays continuously evolved, getting better each time, with technology rapidly advancing after the widespread implantation and publicity of GUIs across many personal devices. This led to a surge in the demand for more attractive, professional, and more eye-catching visuals in the market.

Succeeding was the phase that witnessed the rise of the touchscreen. The touchscreen era opened new ways for users to interact with their devices. GUIs started drawing even more real references to real-life objects and textures. This type of design was known as Skeuomorphism.

The early 2000s saw much more detailed graphics in the department of interface design. 'Icons' and other such elements were being designed in a manner to replicate their real-time counterparts to add a realistic touch. The heavy use of reflections, gradients, and shadows was undertaken to make things pop more. The digital space acutely imitated the real world with the advancements in displays and processing power. A fine example of the above feature would be the flipping of pages on iBooks the same way one could on regular books.

A few years down the line, users began to feel a sense of disenchantment. It was partially since all of digital space was mimicking and recreating our reality and hence, for the end-user, there was not much to look forward to. Things were looking overemphasized and users had begun experiencing a lack of interest in it.

In 2002, the idea of Flat Design was born and it became the game-changer in the field of UI's. It brought about simple, clean, no textures or reflective surfaces, no glass or leather. Skeuomorphism in all its shadowed glory became a thing of the past. By now, users had already become habitual to interfaces and their functionality, and so they had no real need for real references to help them understand how things functioned.

This brought about a dramatic shift from hyper-realistic designs to the new, almost flat design, that was to be seen everywhere. An added advantage of the new flat design was that applications took less time to load as interfaces became less graphic-intensive. In contemporary times, flat design is a mainstream interface design. The flat aesthetic is to be seen everywhere, the websites we surf and mobile apps we so often use.

8.11. WHAT TO EXPECT NEXT?

Design trends come and go. There is no permanency in design and Skeuomorphic design and flat design are no exceptions. Many predictions have been made concerning technology and its advancements in the next few decades and how future interfaces might look like. Let's split the following examples according to current UI trends and potential future UI predictions.

8.11.1. Current UI Trends

VUI's offer the distinct feature of hands-free interaction between the user and the system through voice or speech commands. Examples of VUI's are Siri, Google Assistant, and Alexa. The added advantage that a VUI holds is that it allows for a very natural, life-like interaction that opens up many possibilities. Below are two examples that explain how interfaces are designed for voice.

AR experience is all about the amalgamation of the digital world with the real-time physical world. It offers an ideal, almost flawless interaction in real-time, and these interactions happen in an environment that has a real-world feel to it.

Augmented reality can be experienced in apps like Google translate, Pokémon Go, and Lenskart. While keeping this in mind, the concept of traditional navigation design patterns can be somewhat redundant, and AR will aid in a new era in user interface design. Below are a few examples of AR that shows how UIs are designed for augmented reality.

VR is a computer-generated simulated experience that can be either similar to or completely different from the real world. In this, a person can interact in the realms of an artificial three-dimensional environment.

Applications of virtual reality can include entertainment (e.g., Gaming) and educational purposes (e.g., Medical or Military training).

Currently, these systems use electronic devices such as VR headsets or gloves fitted with sensors, multi-projected environments to generate realistic images, sounds, and other sensations that simulate a user's physical presence in a virtual environment. Brands like Facebook, Microsoft, and Google have ventured into VR and offer these headsets as of now.

8.12. FUTURE UI TRENDS

8.12.1. Wearable User Interfaces

Wearable computers are electronic devices that can be worn easily like an accessory or apparel. They could be in the form of a pair of gloves, eyeglasses, a watch, or even a suit. The major highlight of wearable UI is that it allows for a handfree experience and does not meddle with or disrupt our daily activities.

A few examples of wearable tech that display how UIs are designed for wearables in which your skin might be the touch screen are mentioned below.

Figure 8.7: An illustration of wearable smartwatch.

Source: Image by unsplash.com.

8.12.2. Holographic User Interfaces

A holographic display is a type of display that utilizes light diffraction to create a virtual three-dimensional image of an object. Holograms are unique and distinguished from other forms of 3D imaging in that they do not require the aid of any special glasses or external equipment for a viewer to see the image.

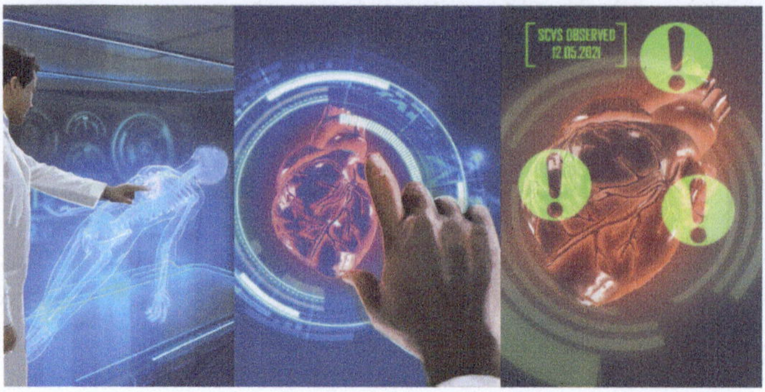

Figure 8.8: An illustration of holographic user interface.

Source: Image by Flickr.

The world of science fiction has provided a great platform for people to express their ideas and fantasies too. Below are a few examples of Holographic projections from the movies Iron Man and Avatar that show how UIs are designed for Hologram.

8.12.3. Autonomous User Interfaces

It would roughly take about 35 years for a fully autonomous vehicle to be accepted fully by society. The foremost reason for this delay is that human beings find it tough to completely trust an automated system.

Yet, what can be certainly assumed and predicted is that semi-autonomous vehicles, which mean intelligent cars with limited autonomous modes, that can handle some situations, but would still require the driver to take the wheel occasionally, will be entering our lives in a full-fledged manner. Brands like Tesla, Volvo, and Waymo & Google have already begun working on these ideas.

An example of semi-autonomous and fully-autonomous vehicles that show how UIs are designed for the automobile industry is mentioned below.

8.12.4. Gesture-Based User Interfaces

Gestural UI involves using specific gestures, like scrolling, pinching, and tapping to operate an interface. It also refers to gesture recognition, tilting, eye motion, and shaking. This technology has come a long way and has evolved from very basic motions and applications to the highly complex ones being used today, and it has now entered our everyday lives. Constant advancements in this technology are ensuring that its future possibilities are also incredibly exciting.

8.13. THE IMPACTS OF UI DESIGN ON EVERYDAY LIFE

Science fiction exercises a major influence on the technological world often by showing us the unattainable. The iPad that is a common sight to behold in today's world was seen long back as a tablet computer in Star Trek: The Next Generation.

Similarly, as is shown in the 1989 cult classic, 'Back to the Future II,' people now do use tablets and computers in their daily life to work and play games with the help of a boost guide they can find online. The man behind the world's first flip phone, the Motorola Star-Tac, was inspired by the Star Trek communicator.

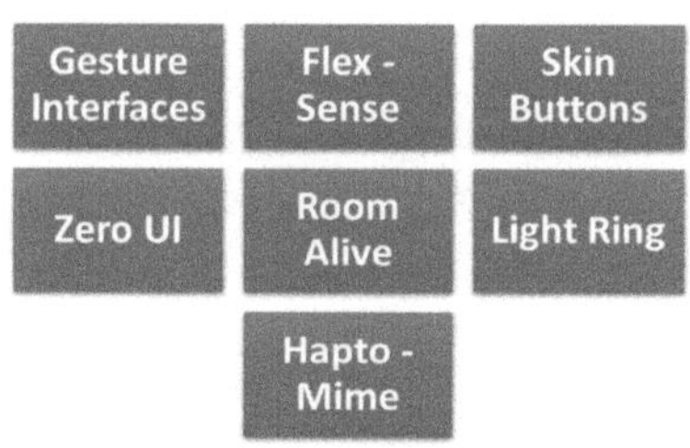

Figure 8.9: The impacts of user interface design on everyday life.

Innovations in technology have promised us that the future is bright and very exciting, and in a way, the future is already here!

Let us take a closer look at some of the examples of how future UI design will impact our daily lives.

8.13.1. Gesture Interfaces

Minority Report and Iron Man are the two films that have to date showcased the most desirable and memorable futuristic user interfaces. These interfaces are the work of inventor John Underkoffler. He believes that the feedback loop between science fiction and reality is growing exponentially with every new summer blockbuster. He goes on to say, "There's an openly symbiotic relationship between science fiction and the technology we use in real life. The interface is the OS – they are one."

8.13.2. Light Ring

Light Ring from Microsoft Research uses infrared to detect finger motion and a gyroscope to determine orientation, and it can turn any surface into an interface. It offers the users to tap, draw, flick and drag on a book, your knee, or the wall. As of now, the interaction is limited to the use of only one finger, yet it offers a really attractive and natural-looking way for user gestures.

Wearable computing has scaled new heights with this latest technology. In the future, controlling a personal device anywhere and in any way suitable to us could be possible with this technology. The nature of using this technology is similar to using a mouse, so the users are already aware of the functional aspect of the product.

8.13.3. Room Alive

Room Alive is Microsoft Research's proof-of-concept prototype and a follow-up to IllumiRoom, which was presented at CES 2012. It transforms any room into an immersive, augmented, magical entertainment experience. Both Room Alive and IllumiRoom are steps towards a "this-is-our-house-now" Kinect future.

The new system goes beyond projection mapping around a TV by adding input-output pixels on top of everything in the room. Room Alive uses multiple depth cameras and spatially mapped projectors to overlay an interactive screen from which there is no escape.

It enables game designers to create projection-mapped games independent of the particular room that the content is displayed. All its content is driven in real-time and dynamically adapts to the exact color and geometry of the user's living room. For instance, it can automatically detect the floors and walls of the room.

Imagine "real-life" video games that transform your living room into the world of the game. Or imagine virtual home decoration, projecting your vision of what you want to rearrange or add to your home's decor.

8.13.4. Skin Buttons

The Skin Buttons project makes use of tiny projectors to display interactive icons on the skin around the watch face. This particular technology is remarkable in the sense that although it expands the interactive zone around a smartwatch, it retains the same size and does not make it bigger physically. Another advantage is its affordability, as the projector parts cost less than $2 and can even offer to increase battery life by shifting the workload from the main display.

8.13.5. Flex-Sense

The FlexSense is a transparent self-sending deformable surface that can reconstruct complex deformations without the need for any external sensing, for instance, cameras. It is a thin film or sheet of plastic, but its embedded piezoelectric sensors are capable of detecting exactly what shape it's in. This allows for all kinds of intuitive, paper-like interactions. For example, flipping up a corner to reveal something underneath, toggling layers in maps or drawings.

Imagine cell phone cases that react as you peel the cover. Or interactive books or children's books that react as you turn a page.

8.13.6. Hapto-Mime

HaptoMime uses ultrasound to create tactile feedback in midair. For example, if you feel like you're touching a hovering image when there's nothing there at all. It's produced by a hidden LCD and an angled transmissive mirror. This technology holds innumerable prospects and possibilities for any kind of public display.

8.13.7. Zero UI

Zero UI isn't a new idea. It is all about using sensory experiences (gestures, movement, and voice) that automatically trigger a reaction from a device. Amazon Echo If you've ever used an Amazon Echo, changed a channel by waving at a Microsoft Kinect, or set up a Nest thermostat, people have already used a device that could be considered part of Goodman's Zero UI thinking.

It's all about getting away from the touchscreen, and interfacing with the devices around us in more natural ways. With methods such as haptics, computer vision, voice control, and artificial intelligence, Zero UI represents a whole new dimension for designers.

Over time, these technologies are becoming inherently natural and more automated for the new crop of users of the current generation. It will bring about a more immersive and seamless computing experience that will continually challenge our capability of storing and processing the vast pool of knowledge they have to share.

The prospects and possibilities for change are both formidable as well as inspiring and are certainly awaited as the doors for new and innovative technologies and groundbreaking products to enter the market are opening.

8.14. THE FUTURE OF UX/UI DESIGN

Great design not only entails great appearances but must also be synonymous with intuition and functionality. If a website or an app is visually appealing but scores low on utility and functionality or is not user-friendly, chances are that it will never gain popularity amongst the users.

Digital design has to be created in a manner that it is of service to its user and human-centered design has to find solutions to overcome all the stumbling blocks that we face today and tomorrow. This explains why user experience (UX) and user interface (UI) design is the central idea of the future of design.

UX design, as the name infers, is concentrated on creating a functional and captivating experience. Research, mock designs, and frameworks are the main constituents of the UX design process. The UX designer builds an operative framework establish on their research and findings and hands it off to the UI design.

The UI designer further continues to refine the frameworks and prototypes to create a high-fidelity deliverable that is later produced by a web or app developer.

Today, around the world, there is a major surge in the demand for UX designers and UI designers. In the cut-throat competitive market that exists today, a company has to possess a great brand experience and solid user experience to survive and thrive. There is no scope for products that offer a poor user experience.

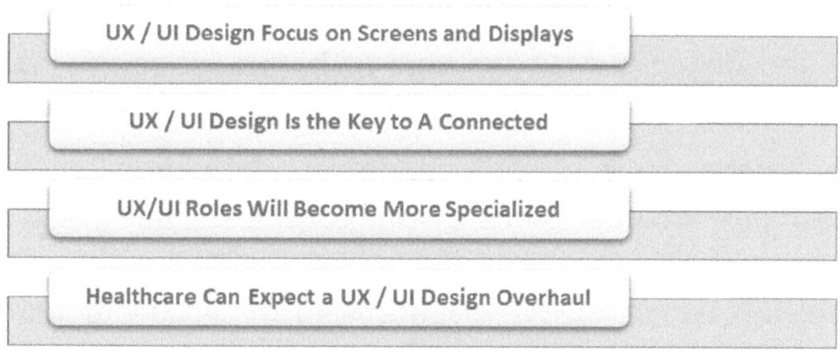

Figure 8.10: The future of UX/UI design.

Design is an ever-evolving area, and thus its future too is highly unpredictable. Yet, its importance will not lessen; it would rather grow as companies look forward to serving the users more than ever before. The future for design is promising, with a strong emphasis on the combination of attractive and appealing design and seamless user experience.

The demand for design is expected to grow exponentially in the coming years, and it is anticipated to expand further with new technologies coming into play. Here are five UX/UI design trends for 2019 and the future of design.

8.14.1. UX/UI Design Will Continue to Focus on Screens and Displays

- Desktop and mobile experiences are not going away;
- UX/UI design needs to create a seamless transition between the two.

That's according to Aaron Fazulak, the director of Design Education at Flatiron School. An estimated 5 billion people use mobile phones in the world, and approximately 2.85 billion of them turn out to be smartphone users. As per the statistics of 2018, Americans, on average, spend 22.5 hours a week online, according to statistics for 2018.

With just over 50% of web traffic coming from mobile, UX/UI designers need to divide their focus equally on both web-based and mobile-based experiences. UX/UI designers will also be faced with the challenge of creating a seamless experience from the desktop to mobile, and vice versa.

8.14.2. UX/UI Design Is the Key to a Connected, Multi-Platform User Experience

- Voice assistants add another interface for users;
- Augmented reality apps require great UX/UI design.

Screens and displays will continue to be prioritized by UX/UI design with a future focus on augmented reality (AR) and new technologies.

Smart assistants have already made our daily lives easier and more efficient and have become a part of our routines. The massive popularity of Siri, Alexa, and Google Assistant is proof of that. The year 2018 saw over 100 million Alexa devices being sold. Apart from that billions of smartphones with Google Assistant, Siri, or Bixby pre-installed were sold too that only adds to their ever-growing demand and popularity.

Though not every iPhone owner needs to be a user is using Siri and the same goes for any Android owner with Google Assistant. Yet, there's a lot of potential with voice assistants. We expect more integration, which means creating a unified user experience.

Take the Amazon experience, for example. "You might ask a question on Alexa that then turns into something on your mobile app and then you're going to hop on a desktop application to order something on Amazon," Aaron says. Connecting all those dots along the journey requires great UX/UI design.

Aaron also has high hopes for AR and expects to see increased UX/UI design demand in AR in the coming years. AR faces the major hurdle of enhancing our everyday life without ever feeling obtrusive and prominent.

To illustrate that point further, we can study Google Maps, which gives us a great example of the need for great UX/UI design with Google Maps. Google launched Google Maps AR in 2018. The key highlight of this edition was that instead of directions on a map, those directions were overlaid on what you were seeing on your smartphone.

Last year's unveiling included a cute fox that helped guide you along the way. Turns out the fox was too cute. Therefore, UI/UX design will continue to hold prominence and more advancement will help popularize AR in the coming years.

8.14.3. UX/UI Roles Will Become More Specialized, But Designers Need to Stay Flexible

- Expect to see specific more roles within UX/UI design;
- UX/UI designers need to adapt quickly as technologies and experiences change.

Up until a few years back, companies would often hire an all-in-one designer. This designer was expected to have a basic working knowledge of UX, some UI, some graphic design, and even some front-end development. Generalists are common in a young field. Over some time, skills and needs become better defined, leading to UX designers and UI designers.

Yet, these roles too would be considered largely broad. Each experience discussed above, be it web-based experiences, mobile-based experiences, voice-based experiences, and AR-based experiences come with its own set of challenges. It is expected of designers to be familiar with each experience, but the result turns out even better if they hold major expertise at one or two experiences.

Specialization has already gained momentum, and it can be seen happening around us. For instance, while browsing through Glassdoor, plenty of opportunities for UX designers and UI designers pop up. Apart from these, roles such as Product Designers, UX Researchers, Visual Designers, Web Designers, UI Engineers, and Interaction Designers can also be seen. Thus, it can be easily concluded that as new technologies mature and develop many more specializations will crop up.

Since experiences are fluid, it is necessitated for UX/UI designers to comprehend the very technique of communicating with all the tools and technologies. "Building multiple skill sets as a designer is increasingly important," Aaron says.

Possessing great soft skills is a pre-requisite for a UX/UI designer as a major part of their job requires them to communicate across multiple teams. If observed on a minute level, right from the actual user to the engineer, humans are at the very core of UX/UI design, and therefore, the ability to hear out, connect with, and communicate their needs is of paramount importance to creating a meaningful and functional experience.

With rapid advancements in technologies, the tools and languages used to create new experiences also undergo major transformations. "You need to be able to jump from one programming language, or one prototyping tool, to another, and we're teaching our students to learn and to be excited about the changes that are coming," Aaron says.

Those changes include whole industries that need to improve their user experience.

8.14.4. Healthcare Can Expect a UX/UI Design Overhaul

- Increased demand for UX/UI designers in healthcare;
- Direct-to-consumer healthcare startups that serve the user are growing rapidly.

UX/UI design is generally linked to tech or creative industries. But if we ponder a little more, we are reminded of all the harrowing experiences that have caused trouble for us the most this year. For instance, while browsing through multiple portals, we often feel exhausted or are left utterly confused by the hard-to-navigate menus. These industries need to remodel themselves and quickly adapt to the digital world and need UX/UI designers to help make that happen.

Of late, that demand is on the rise in healthcare. Telemedicine, for example, is a fast-growing sector in healthcare. Teladoc or Talkspace, for example, make it easy to connect to a doctor or a licensed therapist.

The birth of Direct-to-consumer healthcare has also pushed for UX/UI design demand in the industry. Brands like Oscar, Candid, Hims, Quip, Cove, Hubble, and others are creating intuitive experiences to improve access to medicine.

Hospitals and health insurance companies can use UI/UX designs to reap financial rewards as also benefit from a better user experience. Healthcare ranks among the slowest sectors to adopt digital innovation. A boom in digital innovation can lead to higher customer satisfaction and lowered infrastructure costs for the insurance company. An intuitive app that includes a paperless billing option with an auto-pay function would make for a win-win situation for both the user and the company.

Another advantage of UX/UI design could be seen if it is used for internal software. This would greatly benefit the employees. Internal tools often cause more problems than they solve. While internal tools are essential, most companies don't make them a priority. Improvements in the UX/UI design of the internal tool help the company run soundly and also helps keep employees happy.

8.14.5. What's On the Inside Counts Too

- Well-designed internal tools are needed;
- Employee satisfaction is as important as customer satisfaction.

Employees are often found using the internal tools provided by their organization which was built solely for functionality. These tools are neither intuitive nor pleasing to the eye. They are built to just get the job done.

A quick Google search will prove this point as one can find plenty of articles detailing subpar internal tools. A web app connecting you to customers may be missing key features like a search function. Or scripts meant to automate tasks have poor documentation.

These experiences are disappointing for employees who just want to do their job efficiently. As a result of that, it so happens that an internal tool that's trying to do too much is built by a small team who were lacking on time and resources to refine the process.

Another situation that often arises is that new processes keep accumulating over the years. Technical resources are erratically built upon others. In many ways, it's a house of cards where nobody wants to try and fix it because they're scared it will collapse, and so it's left unattended.

A UX designer or a UI designer would come in handy in such situations. They can streamline that process for the smooth running of the company and create better tools for employees. As a result that, employees will experience heightened productivity and lower frustration levels.

User interfaces—the way we interact with our technologies—have evolved a lot over the years.

Right from the beginning, with the original punch cards and printouts to monitors, mouses, and keyboards, all the way to the trackpad, voice recognition, and interfaces designed to make it easier for the disabled to use computers, interfaces have come a long way and have grown in leaps and bounds in the last few decades.

But there's still a long way to go, and there are many possible directions that future interface designs could take. A lot of directions have already been seen, and they promise an enthralling experience of change in our lives.

8.15. BRAIN-COMPUTER INTERFACE

In a brain-computer interface, a computer is controlled purely by thought (or, more accurately, brain waves). It is being carried out by a variety of approaches, including direct brain implants, full helmets, and headbands that capture and interpret brain waves.

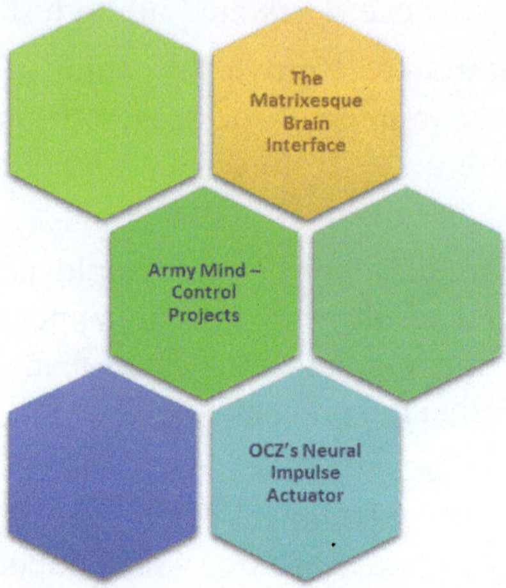

Figure 8.11: Brain-computer interface.

8.15.1. Army Mind Control Projects

According to an article in Time from September 2008, the American Army is actively pursuing "thought helmets" that could someday lead to secure mind-to-mind communication between soldiers. According to the article, the goal is to achieve a system where entire military systems could be controlled by thought alone.

Though it can be agreed that this level of technology is still far off, the fact that the military has awarded a $4 million contract to a team of scientists from the University of California at Irvine, Carnegie Mellon University, and the University of Maryland, it certainly tells us that we might be seeing prototypes of these systems within the next decade.

8.15.2. The Matrixesque Brain Interface: MEMS-Based Robotic Probe

Researchers at Caltech are working on a MEMS-based robotic probe that is capable of implanting electrodes into the brain to interface with particular neurons. While it sounds very The Matrix-y, the idea is that it could allow for advanced control of prosthetic limbs or similar body control.

They have achieved success in designing the software part of the device through the micro-mechanical part (the part that goes into your brain) is still under development.

8.15.3. OCZ's Neural Impulse Actuator

The NIA is a headband and controller that integrates an electro-myogram, an electroencephalogram, and an electrooculogram to help permit it to translate eye movements, facial muscle movements, and brain waves. The unique feature of the NIA is that it can be set up to work with virtually any game; the controller simply translates the input into keystrokes.

8.16. BIOMETRIC AND CYBERNETIC INTERFACES

In computing, cybernetics most often refers to robotic systems and control and command of those systems. Biometrics, on the other hand, refers to biological markers that every human being (and all life forms) has and that are generally unique to each person. These are most often put to use for security purposes, such as fingerprint or retina scanners.

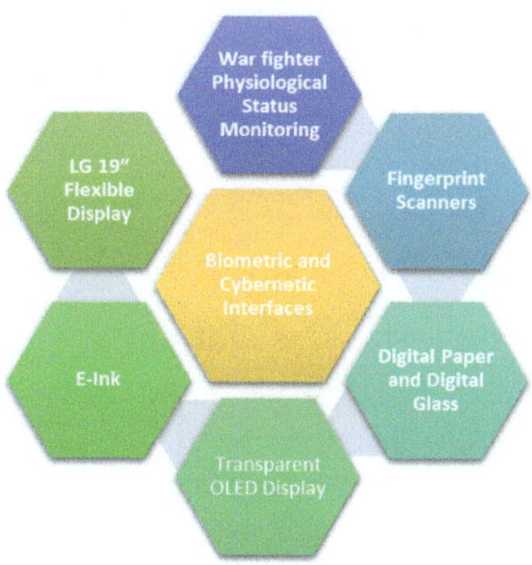

Figure 8.12: Biometric and cybernetic interface

Let us have a look at a few current biometric and cybernetic interface projects.

8.16.1. Warfighter Physiological Status Monitoring

The Military Operational Medicine Research Program is developing sensors that can be embedded into clothing to monitor soldiers' physiological well-being. Monitoring real-time health is the major goal of these. They would also input additional variables into predictive models the military employs for the assessment of the likely success of its missions.

8.16.2. Fingerprint Scanners

For a long time now, fingerprint and hand scanners have been showcased in movies as high-tech security devices. They are now easily accessible to all within the past few years.

Most commonly used to either allow or deny entry or access to certain specific users for a computer system, vehicle, or a high-security area. Fingerprints being unique, offer a full-proof way of ascertaining who gaining access to what and when.

8.16.3. Digital Paper and Digital Glass

Digital paper is a flexible, reflective type of display that uses no backlighting and simulates real paper quite well. The major advantage of digital paper is that mostly it does not need any power except when changing what it's displaying. Thus, the devices that use it have very long battery life. Digital glass, on the other hand, is a transparent display that otherwise resembles a standard LCD monitor.

8.16.4. Transparent OLED Display

In the year 2010, Samsung showcased a prototype of a new, transparent OLED display on a notebook. The display is unlikely to appear in notebooks in its finished form, but it might be used in MP3 players or advertising displays in the future, according to the company.

8.16.5. LG 19″ Flexible Display

It is often predicted that Flexible e-paper displays might replace paper one day. in sharp contrast to their stiff and counterparts, e-paper can be nearly as flexible as real paper (or card stock, at least), and almost as thin. LG has created a 19″ e-paper display that's flexible and made of metal foil so it will always return to its original shape. It is predicted that this particular type of display will be hugely popular for reading newspapers or other large-format content in the future.

8.16.6. E-Ink

E-Ink technology is a captivating technology that has many interesting implications in the packaging and media industries. E-ink is a proprietary paper technology that's already seen some real-world use (such as on the Esquire cover from October 2008). Currently, it is only available in grayscale, but soon it will be available in color too.

All popular eBooks including the Kindle, Barnes & Noble's Nook, and the Sony Reader use the e-ink technology to provide an appearance similar to that of a printed page. Though as of now, it has only been put to use within a rigid display, in the future it could be used otherwise.

8.17. TELEPRESENCE

Telepresence consists of the remote control of a drone or robot. They are now a common sight to behold and are mostly seen in the scientific and defense sectors, and are being used for vastly different purposes.

While in some cases, only visual input is offered by these devices but in others (such as medical telepresence devices), a more complete simulation is created. Below are some of the best telepresence projects currently underway.

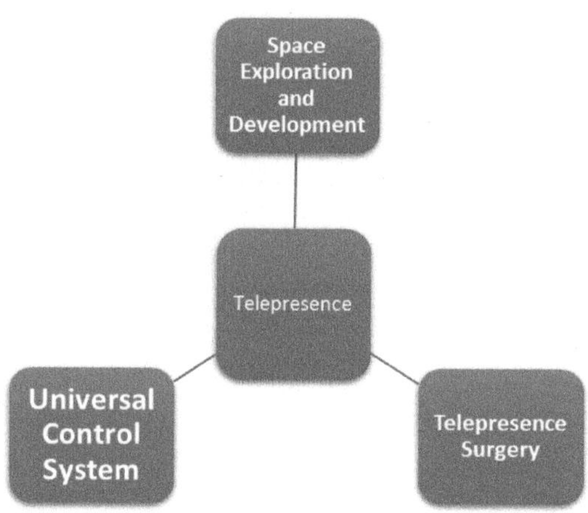

Figure 8.13: Telepresence.

8.17.1. Telepresence Surgery

Surgeries that require minimal invasion can now be conducted via telepresence. In such cases, the surgery is performed by a robot on the actual patient, while a surgeon controls it via a remote. This type of surgery is found to give better results than manually undertaken surgeries that require the use of long, fulcrum instruments.

This technology uses a combination of telerobotic, sensory devices, stereo imaging, and video and telecommunications. This in turn offers the surgeons, a complete sensory experience of traditional surgery. In addition to that, surgeons also receive real-time feedback, including the finer details such as the pressure they would feel when making an incision in a hands-on surgery.

8.17.2. Universal Control System

The Universal Control System is a system developed by Raytheon (a defense contractor) for directing aerial military drones. The interface somewhat replicates a video game, with multiple monitors to give operators a 120° view of what the drone sees.

Raytheon closely observed the existing technology that the drone pilots were using (which consisted of standard computer systems—a nose-mounted camera and a keyboard) and concluded that betters systems were being used by gamers. This set them off to develop a drone operating system based on civilian business games (and even hired game developers). Augmented reality and other futuristic interface elements are also included in the finished system.

8.17.3. Space Exploration and Development

The innovative technology of telepresence could be used to allow humans to experience space environments from the safety of earth. This technology could enable us to remotely traverse distant planets from our planet.

It is facing a few challenges in its way; the biggest being the delays in communications over a long distance. There has been progressing in that area, and in the coming years, it might overcome that hurdle and allow for remote exploration.

8.18. THE FUTURE UI TREND OF 2025

Once we find something that works, we get used to it, and we crave the next big thing. The next innovation. When will the future finally arrive? And when it does, how long will it take us to get used to it, and eventually be bored with it?

Let's jump five years ahead and try to predict where UI design will be then based on what we know now.

8.18.1. One style to rule them all?

With their recent WWDC announcements, Apple has cemented the rule of a variant of the popular Modern/Soft UI style. As always, they had their twist on it, and it did become a little controversial. The main issues were excessive usage of skeuomorphic elements and certain alignment issues with iOS home screen widgets.

The UI revolution (in the futuristic, touch-enabled sense) started in 2007 with skeuomorphism, because touch interfaces with that fidelity were in their nascent stage and scary.

They required people to memorize the patterns and make sense of the possibilities. That explains why the virtual books in those days were on a shelf, and the notepads were stitched leather and a paper texture.

More familiarity with the UI's lead to the "modernization" phase. Textures were slowly removed, except for the note's app, which to this day has a barely visible paper texture under the text.

Corners were more rounded than in the previous versions. The target audience of touch interface too saw a gradual shift and grew from early adopters and the tech-savvy to "the general population."

In the present day and age, the general public is using the same apps as Bill Gates or Elon Musk. People like the rounded corners, readable sans-serifs, and a splash of color here and there. Aside from the lack of apps, Windows Phone UI may have been a bit too revolutionary for its good.

8.19. MODERN UI 2020

The explosion of the modern UI trend is indisputable. Both the new iOS, Material Design, and what you can see on Dribble use a lot of the same patterns, because of the familiarity with these patterns and knowledge of how they work. People prefer them and can easily comprehend them.

Despite having differences and variances in their characteristics, overall, they are a part of "the same" trend.

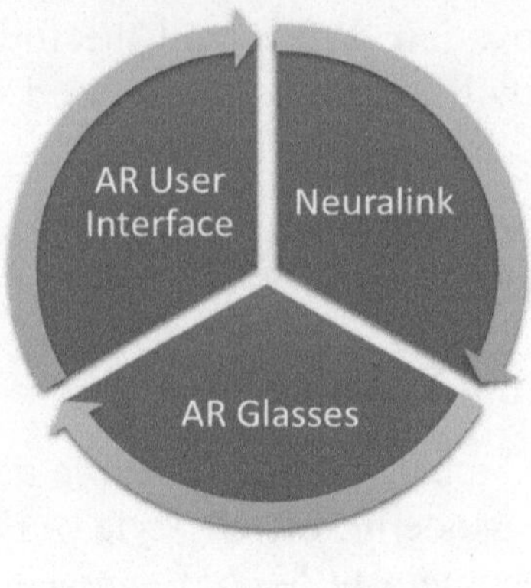

Figure 8.14: Modern user interface 2020.

Neumorphism offered a brand new and exciting trend and was lapped up almost immediately by users and went viral despite all of its controversies and poor-quality design choices. It was introduced at a time when users had realized that almost all UIs are similar in appearance and were craving for a change.

Yet Neumorphism does not hold a promising future and will soon be treated as a non-serious novelty. In some cases, some neuromorphic elements may end up merging with the Modern UI for that distinct style.

8.19.1. Neuralink

Brain-computer interfaces are beyond theory now, with significant progress year by year. Elon Musk famously said at Joe Rogan's that he believes in more widespread adoption in the next decade. One of the many hurdles this technology is faced with is that it has to catch up with the decreasing fear of having brain implants. In general, the world is not yet equipped to handle such a situation.

8.19.2. AR Glasses

Apple has been reportedly making steady progress on their set of AR Glasses for quite some time now. It is anticipated to turn out to be a landmark moment for UI design. Its popularity and how consumers react to it will determine the future of AR glasses. We know the pros use Google Glass or Microsoft

HoloLens, but most real people wouldn't go for a coffee in a dorky-looking headset just to scroll Instagram with their eyes easier.

In case Apple succeeds and creates a pair of glasses (or contact lenses eventually) that will look "normal" or even fashionable it might bring about a revolution in the world of AR.

8.19.3. AR UI?

In this case, the UI will still use the Modern trend as its starting point, but it would require a major adjustment for one crucial element. Generally, our UI was displayed on a screen. A flat, glass canvas that turns black when switched off.

It's only a window to the amazing new worlds when it's on. Glasses are different because you see through them. But in this case, the canvas is no longer a dark glass panel, but rather the entire world. It certainly raises a few questions as it was still quite difficult to ascertain the background on which the UI would be on display.

8.20. CONCLUSION

In the conclusion of the chapter, this chapter discussed the future of user interface design. This chapter also discussed the current state of user interface design, as well as the rise of voice user interface and accessibility. In this chapter, it has been discussed how artificial intelligence will disrupt user interface design. This chapter discussed the various trends and prophecies of user interface design in the future. This chapter also discussed the past, present, and future scenarios of the user interface.

Towards the end of the chapter, various future trends of the user interface have been discussed which include a wearable user interface, holographic user interface, autonomous user interface, gesture-based user interface. This chapter also discussed the various impacts of the user interface design on everyday life such as gesture interfaces, light ring, room alive, skin buttons, flex–sense, Hapto–mime and zero UI.

This chapter discussed the future of UX or UI design, that it will continue to focus on screens and displays, UX/UI design is the key to a connected, multi-platform user experience, it will become more specialized, but designers need to stay flexible, and healthcare can expect a UX/UI design overhaul. In this chapter, the concept of biometric and cybernetic interfaces has been discussed, such as warfighter physiological status monitoring, fingerprint scanners, digital paper, and digital glass, transparent OLED displays, and E-ink.

REFERENCES

1. Carl, B., (n.d). *The Future of UI Design and Its Impacts on Everyday Life*. Wrecking Ball. Studio + Labs. [online] Wrecking ball. STUDIO + LABS. Available at: https://www.getwrecked.com/the-future-of-ui-design-and-how-it-will-impact-our-everday-lives/ (accessed on 6 July 2021).
2. Craig, W., (n.d). *The Future of User Interfaces*. [online] WebFX Blog. Available at: https://www.webfx.com/blog/web-design/the-future-of-user-interfaces/ (accessed on 6 July 2021).
3. Eric, B., (2020). *What's the Future of UI Design? Trends and Prophecies*. [online] Careerfoundry.com. Available at: https://careerfoundry.com/en/blog/ui-design/whats-the-future-of-ui-design-prophecies-and-trends-for-2018-and-beyond/ (accessed on 6 July 2021).
4. Flatiron School, (2019). *The Future of UX/UI Design*. [online] Flatironschool.com. Available at: https://flatironschool.com/blog/future-ux-ui-design (accessed on 6 July 2021).
5. Graham, C., (2018). *Codefirst: The Future of UI Design*. [online] InfoQ. Available at: https://www.infoq.com/articles/codefirst-future-ui-design/#:~:text=Data%20Will%20Drive%20the%20Future,scientific%20like%20Google's%20material%20design (accessed on 6 July 2021).
6. Michal, M., (2020). *The Future UI Trend of 2025?* [online] Medium. Available at: https://uxdesign.cc/the-future-ui-trend-of-2025–14d9fdf6745 (accessed on 6 July 2021).
7. Omkar, C., (2020). *The Past, Present and Future of User Interfaces*. [online] Medium. Available at: https://medium.com/17seven/the-past-present-and-future-of-user-interfaces-2df669dc63e3 (accessed on 6 July 2021).

Index